INTERESTING FACTS

FOR UNLEASHED CURIOSITY

50 IDEAS AND ACCIDENTS IN SCIENTIFIC DISCOVERIES

FOR CURIOUS MINDS THAT CHANGED OUR LIVES

DEDICATION

To you, curious soul and tireless seeker of the unknown, who find magic in the hum of a circuit, the glow of the sun on a leaf or the elegant swirl of water making its way through a maze. You who marvel at the unexpected and delight in the mysteries of our everyday world.

As a fellow explorer of the hidden corners of science and the wonders of technology, I invite you to discover the surprising and transformative stories that have quietly shaped our modern world.

Thank you for joining me on this journey into the sparks of genius that connect us to the wonders of our world.

O. Bennet

Copyright © 2024 Ely & Oly books
All rights reserved.

Prologue

Everything usually starts with a question. Not the kind of question you find in textbooks, but something stranger, something whispered in the quiet moments of curiosity. Can bacteria breathe metal? Can a turtle think? Can quantum mechanics guide a bird across continents? These are not the questions of conventional science. Yet, as history has shown, the most outlandish questions often lead to the most brilliant revelations.

This book is a journey into the hidden sparks of genius: the serendipitous, accidental, and downright bizarre moments that forever altered the trajectory of science and technology. You'll meet the unlikely heroes of discovery: microbes that turned mud into electricity, liquids that solved problems with fascinating elegance, and quantum phenomena that bring to life the deepest mysteries of biology. These are not only stories of triumph, but also of failure, chance and audacity, stories that remind us that breakthroughs often arise where no one thought to look for them.

We will walk through muddy riverbeds where electric bacteria sparked the renewable energy revolution. We'll peer into glass tanks where swirling fluids revealed solutions to puzzles no supercomputer could solve. And we will take to the skies with migratory birds guided by the invisible hand of quantum mechanics, their trajectories etched in starlight. Each story is a window into the hidden corners of scientific exploration, where the ordinary becomes extraordinary.

This is not just a book about discovery, but a celebration of curiosity. It is a reminder that the world is full of mysteries that await those who dare to ask questions and follow the threads, however strange or unexpected they may seem.

As you turn these pages, let your imagination run wild. For science is not just the pursuit of knowledge: it is the art of wonder, the craft of exploration, and the unwavering belief that even the wildest idea can have the power to change the world.

Contents

Prologue	3
001 Alchemist's Glow	7
002 The Birth of Probability	11
003 Microscope's Hidden World	15
004 The Weight of Air	19
005 Accidentally Magnetic	23
006 Spontaneous Generation's Fall	27
007 Seeing Sound	31
008 Serendipitous Gas	35
009 Mysterious Magnetic Poles	39
010 Taming Electricity	43
011 Galvani's Frogs	47
012 The Accidental Battery	51
013 Cyanotype and Sunlight	55
014 Lavoisier's Elements	59
015 The Whispering Arches	63
016 Penicillin's Accidental Birth	67
017 X-Rays: Vision Beyond Sight	71
018 Seeing the Double Helix	75
019 The Pacemaker's Pulse	79
020 Silkworm's Steel	83
021 Discovering the Electron	87
022 Plastic's First Step	91
023 The Microwave Oven Mistake	95
024 Radar for Rain	99
025 The Velcro Hike	103
026 Neurons on Film	107
027 Bubble Wrap Blunder	111
028 Super Glue's Sticky Situation	115
029 Plastic Surgery's Battle Origins	119
030 Magnetic Resonance	123
031 The Laser Leap	127

032 Silicon Valley's Seeds 131
033 The Integrated Circuit 135
034 LED's First Light 139
035 Microprocessor 143
036 The Internet's First Nodes 147
037 Polymerase Chain Reaction 151
038 Quantum Dots 155
039 Cochlear Implants 159
040 The bionic eye 163
041 The Unbreakable Code 167
042 The Accidental Discovery of Fullerenes 171
043 The Telepathic Frogs 175
044 The Shrinking Machine 179
045 Artificial Earthquakes 183
046 The Cybernetic Turtle 187
047 Plants That Power 191
048 The Bacterial Battery 195
049 Liquid Computers 199
050 Quantum Biology 203
Bibliography 206
Final note of thanks 209
About the author 210

Alchemist's Glow
The First Chemical Luminescence and Phosphorus

In the 1600s, Europe was steeped in **mysticism** and **alchemy**, with scholars and enthusiasts alike driven by a desire to uncover hidden secrets of the natural world. Among them was **Hennig Brand**, a German alchemist obsessed with finding the **philosopher's stone**—a legendary substance that could turn base metals into gold. At a time when empirical science was only just beginning to emerge, alchemists like Brand employed mysterious, often bizarre techniques in their experiments. Alchemy was a mixture of mysticism and early chemistry, and although few truly believed in magical transmutations, the search for "purified elements" captured imaginations and encouraged strange yet fascinating experiments.

Brand's pursuit of the philosopher's stone led him to experiment with **human urine**, a substance he thought might contain a distillable essence of life. He collected urine from the local soldiers, as he assumed that their robust constitution might enhance its properties. Brand believed that if he boiled and distilled the liquid repeatedly, he might isolate a pure, transformative material. After hours of heating, distilling, and condensing, Brand observed an unusual phenomenon: a soft glow emanated from the final product, an eerie, faint light that seemed to pulse in the dark. Although he hadn't found gold, he'd isolated **phosphorus**, one of the first chemically created elements to exhibit **luminescence**—a glow without heat. In this dim yet persistent light, Brand saw something extraordinary, though he couldn't yet grasp its scientific significance.

Brand's discovery was accidental, yet it marked a historic moment. Phosphorus was the first element isolated through a chemical process, and its strange luminescence was unlike anything known at the time. Word of the "alchemist's light" spread quickly, sparking a blend of wonder and fear among the elite circles of Europe. Although phosphorus didn't fulfill the alchemical dream of transmuting metals, it brought attention to an intriguing natural phenomenon. Brand kept his method secret, hoping he might yet profit from his glowing substance, but other alchemists and natural philosophers were drawn to the mystery of this new material.

Importance and Impact

While Brand's phosphorus wasn't gold, it was the beginning of **modern chemistry**. The isolation of an element through chemical reactions, rather than mystical incantations or nonscientific methods, was a crucial milestone. Over time, the use of chemical processes to isolate elements laid the foundation for **elemental chemistry**. Phosphorus would be the first of many discoveries that shifted alchemy from mystical pursuits to more systematic, observational experiments—transforming it into the empirical science that we now call chemistry. Brand's accidental discovery of phosphorus revealed a phenomenon that scientists would soon call **chemiluminescence**, the ability of certain substances to emit light as a result of a chemical reaction.

In addition to its impact on chemistry, phosphorus itself became highly valuable. Due to its luminescent quality, phosphorus was quickly incorporated into **medicinal** and **practical** uses. It was found to be a powerful ingredient for making **matches**, as its glow-in-the-dark properties allowed for more efficient fire-starting, particularly for soldiers and travelers. Its glowing quality also made it useful in early military applications as a signal light in the dark, long before electricity became widespread. Phosphorus brought about a new fascination with elements and compounds, fueling future experiments that led to the identification of other important substances, including sulfuric acid and hydrochloric acid.

Brand's experiment with phosphorus inspired later generations to think differently about **chemistry** and **elements**. Scientists began to consider that materials might be broken down into fundamental components, an idea that was pivotal in the development of the **periodic table** by Dmitri Mendeleev in the 19th century. Phosphorus itself would later be recognized as essential in **biological processes**, especially as a component of DNA and cellular energy molecules like ATP, though such applications were unknown to Brand. Despite being shrouded in mystery and a bit of superstition, the glow of phosphorus opened doors to studies that ultimately demystified the nature of matter and set the stage for modern elemental research.

Chapter Summary

Hennig Brand's accidental isolation of phosphorus in his pursuit of the philosopher's stone may not have led to gold, but it unveiled a new path in science. Phosphorus was the first element to be isolated through chemical means, displaying an eerie glow that captured the curiosity of the European elite and set the foundation for early chemistry. This luminescent element played a critical role in shifting alchemy from mystical interpretations to experimental science, inspiring the systematic study of elements and their properties. The discovery not only marked a move toward modern scientific methods but also demonstrated the practical value of phosphorus in everything from matchsticks to signaling, foreshadowing its essential role in later biochemical research. Brand's discovery was one of the earliest examples of a chance experiment leading to a lasting impact, shaping the world of science in ways he could never have foreseen.

Next Chapter

With the successful isolation of phosphorus, European alchemists and scientists began to reconsider the natural world as a composition of elements that could be systematically identified, categorized, and understood. This shift marked a move toward **experimental chemistry** and the quest to understand matter on a fundamental level. In the next chapter, we will explore the work of **Blaise Pascal** and his unexpected contributions to mathematics, specifically in the realm of probability. Originally intrigued by games of chance, Pascal's research into randomness and outcomes led to the creation of **probability theory**, a cornerstone of modern science that would later influence fields as varied as physics, economics, and artificial intelligence.

002 The Birth of Probability
Gambling and the Dawn of Statistics

In the mid-17th century, Europe was on the cusp of the **Age of Enlightenment**, a period marked by rapid advances in science, mathematics, and philosophy. While most scientific achievements of the era were inspired by practical needs or natural curiosity, one of the most profound breakthroughs in mathematics had surprisingly humble beginnings—in the gambling halls and games of chance enjoyed by Europe's social elite. During this time, dice games and card games were popular among the wealthy, particularly in France and Italy. These games, however, posed a unique problem for players and observers alike: how could one calculate the likelihood of a particular outcome? Questions such as "What are the odds of rolling a seven with two dice?" seemed straightforward but were impossible to answer using existing mathematical knowledge.

Into this environment stepped **Blaise Pascal**, a brilliant young mathematician and philosopher from France. Pascal was fascinated by the puzzles of chance that arose in these games, viewing them as intellectual challenges rather than mere entertainment. In 1654, he received a letter from **Chevalier de Méré**, an accomplished gambler, who was troubled by a specific problem regarding the likelihood of winning certain dice games. De Méré's question was deceptively simple: why was it that some betting strategies appeared to fail even when they seemed mathematically sound? Pascal, intrigued by this problem, began a correspondence with **Pierre de Fermat**, another celebrated mathematician. Together, through a series of letters, they began to devise mathematical approaches to calculate probabilities, unknowingly laying the groundwork for a new branch of mathematics.

Pascal and Fermat's correspondence led them to formulate the **basic principles of probability theory**, including the concept of expected value. Although the practical application at the time was limited to games of chance, their work would ultimately create a foundation for statistical science. By breaking down complex problems into sequences of possible outcomes, they were able to derive probabilities for events that seemed random, offering a logical structure for dealing with uncertainty. Their work was unprecedented, for up until this point, **uncertainty had been considered an obstacle to be avoided**, not an area to be studied. The correspondence between Pascal and Fermat not only provided de Méré with insights into his gambling strategy but also became the basis for the mathematical study of **risk, probability, and decision-making**.

Importance and Impact

Pascal and Fermat's development of probability theory revolutionized the way people approached uncertainty and risk. Initially, their work had a relatively limited reach, influencing only a small circle of mathematicians and gamblers. But as the **concept of probability** began to gain traction, scholars and intellectuals recognized its broader implications. The study of probability introduced a structured way to assess and manage risks, fundamentally changing fields as varied as **insurance, economics, medicine, and engineering**. By quantifying the likelihood of certain events, probability theory enabled a more scientific approach to predicting and managing outcomes, allowing society to make more informed decisions even in uncertain situations.

In particular, probability theory paved the way for **insurance industries** to calculate risks associated with health, property, and life insurance policies. Armed with statistical data and a mathematical understanding of risk, insurers could now develop policies with premiums accurately tailored to individuals' circumstances. Probability also entered **medical research**, where it provided a framework for understanding health risks and treatment outcomes, directly impacting public health and the effectiveness of medical interventions. Probability calculations soon underpinned **actuarial science**, allowing professionals to make financial forecasts based on projected probabilities—a practice critical to modern finance.

In addition to transforming applied fields, the development of probability theory had profound philosophical implications. Pascal himself would go on to apply the concept of probability in his famous **Pascal's Wager**, a philosophical argument regarding belief in God that used probability to discuss decision-making under uncertainty. Probability theory thus introduced new ways to approach not only physical risks but also metaphysical and philosophical questions, offering a quantitative lens through which people could assess choices in life's uncertainties.

Probability's role in everyday life continued to expand, reaching into **psychology and behavioral science** to analyze how humans perceive risk and make decisions. Today, our understanding of probability influences everything from **public policy** and **economics** to **weather forecasting** and **sports analytics**. From its modest beginnings as a response to questions of gambling, probability theory became a cornerstone of modern science, demonstrating how mathematical tools could be applied far beyond their original context. The concepts of probability have empowered societies to take calculated risks, innovate, and pursue progress without being completely paralyzed by uncertainty.

Chapter Summary

The unlikely beginnings of probability theory in gambling halls and games of chance led to one of mathematics' most transformative breakthroughs. Pascal and Fermat's exploration of games of chance, spurred by Chevalier de Méré's questions, gave rise to a mathematical framework that would eventually underpin everything from insurance to economics, medicine to philosophy. The study of probability not only reshaped practical fields by offering tools to measure and manage risk but also shifted philosophical thought, allowing humanity to grapple more effectively with uncertainty. This chapter captures the serendipitous origin of probability and its profound impact on nearly every scientific and societal field that followed. By quantifying risk, probability has allowed humankind to approach the unknown with structured insight, marking an intellectual shift that would influence countless future discoveries.

Next Chapter

As mathematicians began to explore new methods of understanding the world around them, another **revolutionary instrument** was being developed to peer into previously invisible realms. The next chapter will introduce the microscope, an instrument crafted by a Dutch cloth merchant that revealed an unseen world teeming with life. This unexpected discovery would bring humanity closer to understanding the building blocks of living organisms, paving the way for breakthroughs in microbiology and medicine.

Microscope's Hidden World

A Draper's Obsession Opens a New World

In the late 17th century, few people could imagine the existence of a world beyond what the eye could see. **Antoni van Leeuwenhoek**, a Dutch draper and lens grinder with no formal scientific training, was about to change that. Living in Delft, van Leeuwenhoek initially crafted lenses to inspect the quality of fabrics he sold. But as he honed his skill in lens-making, he became increasingly fascinated by the idea of seeing objects more closely than anyone had ever seen before. His curiosity led him to build small, powerful microscopes capable of much higher magnification than any others of the time. Unlike the compound microscopes being developed elsewhere, van Leeuwenhoek's simple, single-lens microscopes reached extraordinary levels of clarity and detail.

Van Leeuwenhoek began examining everyday materials: threads, grains of sand, drops of pond water. What he observed left him spellbound. In water droplets, he discovered tiny, wriggling forms—what he described as "animalcules." He was looking at the first recorded observations of **microorganisms**. Van Leeuwenhoek's excitement grew as he realized he had stumbled upon a hidden world teeming with life, entirely invisible to the naked eye. His discoveries, which he painstakingly documented in letters to the **Royal Society of London**, astounded scientists and laypeople alike. However, as an outsider without formal training, van Leeuwenhoek's observations were initially met with skepticism, and it took years for the scientific community to fully appreciate the significance of his work.

Despite the doubts, van Leeuwenhoek continued his microscopic explorations. Over several decades, he became the first person to observe **bacteria**, **red blood cells**, **muscle fibers**, and **sperm cells**. Each new discovery added to his reputation, gradually gaining him the respect and admiration of established scientists across Europe. Van Leeuwenhoek's observations challenged traditional views of life, sparking debates on **spontaneous generation** and laying the groundwork for future biological research. His contributions were revolutionary, not only in their content but in their insistence that careful observation and empirical evidence could reveal truths unknown even to learned scholars.

Importance and Impact

The significance of van Leeuwenhoek's discoveries cannot be overstated. He essentially opened the door to **microbiology**, a field that would ultimately transform our understanding of health, disease, and life itself. By discovering **microorganisms**, van Leeuwenhoek gave humanity a glimpse into the unseen life forms responsible for many natural processes, including **decomposition** and the **spread of diseases**. His observations would inspire generations of scientists to study microorganisms in greater depth, leading to breakthroughs in medicine, agriculture, and ecology. Although he did not fully understand what he saw, his meticulous records of these "animalcules" provided future scientists with a foundation for studying microorganisms' roles in health and disease.

Van Leeuwenhoek's findings also challenged the **dominant medical theories** of his time, which held that diseases arose from imbalances within the body or from supernatural forces. By revealing the existence of tiny living entities, he hinted at the microbial basis of disease—an idea that would later be solidified in the **germ theory** of disease. His work, though not fully appreciated during his lifetime, foreshadowed the eventual realization that invisible agents could cause infections, epidemics, and even plagues. This insight would transform medicine and public health, leading to the development of **antiseptics**, **antibiotics**, and modern methods of sterilization, all of which have saved countless lives.

The impact of van Leeuwenhoek's discoveries went beyond microbiology. By pushing the boundaries of what was observable, he laid the philosophical foundation for modern science, emphasizing empirical observation over speculation. He demonstrated that curiosity, patience, and a willingness to explore the unknown could yield profound insights into nature. His legacy is a testament to the power of curiosity-driven research, as well as a reminder that significant discoveries can come from unexpected places—even from a draper with a simple lens and a boundless curiosity. Today, his name is synonymous with the microscope, and his contributions serve as a reminder of the vast, hidden worlds that still await discovery.

Chapter Summary

Antoni van Leeuwenhoek's accidental entry into the world of microbiology marked one of history's most profound discoveries. Through his handmade microscopes, he unveiled a universe of microorganisms, challenging traditional beliefs and igniting a field that would forever change human knowledge. His discoveries, initially met with skepticism, eventually won recognition, revealing an unseen world that is now central to biological W. Van Leeuwenhoek's findings laid the groundwork for understanding the microscopic causes of disease, influencing the development of modern medicine and reshaping humanity's understanding of life. His story exemplifies how simple curiosity, even from an untrained mind, can lead to discoveries that transform our grasp of the natural world.

Next Chapter

With van Leeuwenhoek's revelation of a microscopic world, scientists grew increasingly aware of the power of observational tools to uncover hidden truths. As they began exploring other ways to manipulate and examine natural phenomena, a new series of discoveries was on the horizon. In the next chapter, we'll explore the surprising discovery of **atmospheric pressure** and the **vacuum**, as a German scientist named **Otto von Guericke** demonstrated a force so powerful it would shift scientific views of nature and lay the groundwork for both physics and engineering innovations.

The Weight of Air

Magdeburg Hemispheres and the Discovery of Vacuum

In the mid-17th century, scientists and philosophers grappled with fundamental questions about the nature of the physical world, and **vacuum**—the concept of empty space—was a particularly controversial topic. According to **Aristotle**, nature abhorred a vacuum, a belief widely accepted for centuries. This philosophical stance suggested that empty space was impossible, yet the **Scientific Revolution** was beginning to challenge long-held assumptions. Among those questioning the idea was **Otto von Guericke**, a German scientist, engineer, and mayor of Magdeburg. Von Guericke believed that empty space could exist and was determined to prove it experimentally.

In 1654, von Guericke staged a demonstration in Magdeburg that would become legendary. He devised an apparatus known as the **Magdeburg Hemispheres**—two hollow copper hemispheres that fit together to form a sphere. Using a manually operated pump, von Guericke was able to remove the air from inside the sphere, creating a vacuum. To demonstrate the strength of this vacuum, he attached horses to each side of the sphere, instructing them to pull in opposite directions. The crowd watched in amazement as the horses struggled but failed to separate the hemispheres, vividly illustrating the powerful force exerted by the surrounding air.

Von Guericke's demonstration was groundbreaking. It visually communicated the concept of atmospheric pressure—the force exerted by the weight of air pressing down on everything around it. His experiments proved that air itself exerted a tangible weight and that its absence created a unique set of physical conditions. Von Guericke had not only demonstrated the existence of a vacuum but had also showcased its practical implications, hinting at a new understanding of the physical forces shaping the natural world. This experiment marked the beginning of **pneumatics** as a scientific discipline, ultimately challenging the Aristotelian worldview and opening doors to further inquiry.

Importance and Impact

The success of the Magdeburg Hemispheres experiment was a critical turning point in physics and engineering. Von Guericke's demonstration challenged deeply ingrained beliefs about the impossibility of a vacuum, shifting scientific thought away from Aristotle's teachings toward an **empirical**, **experimental** approach. The spectacle of the experiment was profound in its simplicity; it used a powerful, visual demonstration to convey an abstract concept, bridging the gap between theory and observable reality. By showing that air exerted pressure and that this pressure could be manipulated, von Guericke's work laid the foundation for technologies that would later rely on vacuum and pressure principles, including **air pumps**, **engines**, and **barometers**.

The implications of von Guericke's experiments extended beyond engineering. His findings had a direct impact on the field of **astronomy**. If a vacuum could exist on Earth, it suggested the possibility of a vast vacuum in space, leading to a new understanding of the cosmos. This idea would later be pivotal to **Isaac Newton's theories of gravity** and motion, as well as to the development of **space science** centuries later. Von Guericke's experiments were thus not merely a demonstration of atmospheric pressure but a revelation that redefined humanity's perception of the universe and our place within it.

Von Guericke's work also inspired other scientists. **Robert Boyle**, intrigued by the implications of a vacuum, would go on to develop his own vacuum pump and conduct experiments that led to Boyle's Law, which describes the relationship between the pressure and volume of gases. **Boyle's Law** would become fundamental to thermodynamics, establishing a mathematical basis for understanding gases that remains in use today. The development of pneumatics and the ability to create a controlled vacuum environment fueled advancements across numerous fields, including **chemistry**, **engineering**, and **medicine**. Von Guericke's work laid the groundwork for what would eventually become modern **fluid dynamics** and **thermodynamics**.

Furthermore, von Guericke's demonstration influenced the public perception of science. His experiment, conducted as a public spectacle, drew attention to the power of science to unveil hidden truths about the natural world. It also showcased the practical applications of abstract scientific ideas, encouraging a broader appreciation for scientific inquiry. His use of **demonstration science**—making complex ideas accessible and engaging for an audience—would be echoed by future scientists like Michael Faraday and Carl Sagan, who sought to popularize science and make it approachable to the public. The Magdeburg Hemispheres thus became a symbol of the empirical approach, a reminder of science's ability to reshape the world through observation, experimentation, and shared knowledge.

Chapter Summary

Otto von Guericke's Magdeburg Hemispheres experiment marked a fundamental shift in scientific understanding, proving the existence of a vacuum and demonstrating the weight and pressure of air. His public demonstration, with horses pulling in vain to separate the hemispheres, visually communicated the power of atmospheric pressure and challenged Aristotle's longstanding doctrine. The experiment laid the groundwork for pneumatics and inspired future research that would redefine fields such as physics, astronomy, and thermodynamics. By sparking curiosity and attracting public interest, von Guericke's work helped to foster an era where science was seen not only as a pursuit of knowledge but also as a force for tangible progress and exploration. His contribution reverberated through the centuries, influencing scientists like Robert Boyle and establishing a foundation for the study of gases and the development of technology that would harness the power of pressure and vacuum.

Next Chapter

With the concept of a vacuum now firmly established, scientific inquiry would soon turn to understanding the **electrical properties** of materials. In the next chapter, we will explore a discovery made nearly a century later, as early scientists investigated the phenomenon of **natural magnetism**. This curious property of certain stones would spark intense interest and pave the way for breakthroughs in electricity and magnetism. From humble experiments with lodestone to the birth of electromagnetism, this journey would reveal new, unseen forces that govern the world around us.

005 Accidentally Magnetic
The Discovery of Lodestone's Power

The phenomenon of **magnetism** has fascinated humans for centuries, yet its origins were mysterious. One of the first instances of naturally occurring magnetism was discovered in lodestone, a mineral composed of **magnetite** that exhibits a magnetic force strong enough to attract iron. The unique properties of lodestone were observed as early as **ancient Greece** and **China**, where philosophers and scientists noticed how these rocks could pull metallic objects toward them. However, despite their magnetic properties, the underlying causes of this attraction remained largely unknown. Lodestone's influence was so remarkable that it was often regarded with a sense of awe, even **superstition**, believed by some to hold magical or divine powers.

Around the 12th century, lodestone's magnetism began to be used in practical applications, most notably in **navigation**. Chinese inventors were the first to develop the **magnetic compass** using lodestone, capitalizing on its tendency to align itself with the Earth's magnetic poles. This early compass, called a **"south-pointer"**, helped sailors determine direction, providing a revolutionary tool for navigation that significantly advanced maritime exploration. The compass eventually spread westward, reaching the Islamic world and then Europe, where it became essential for explorers embarking on transoceanic voyages. Lodestone's magnetism thus played a crucial role in enabling the **Age of Exploration**, though the scientific understanding of magnetism was still in its infancy.

In the late medieval period, scholars and philosophers in Europe started to investigate lodestone more systematically. Among them was the English scientist **William Gilbert**, who, in the 16th century, conducted rigorous studies of lodestone and other magnetic materials. Gilbert observed that lodestone could **induce magnetism in iron**, leading him to propose that the Earth itself functioned as a massive magnet with its own magnetic field. His seminal work, **De Magnete**, published in 1600, marked one of the first scientific explanations of magnetism and laid the groundwork for future research. Gilbert's studies transformed the perception of lodestone from a mystical object to a scientific tool, setting the stage for further exploration of electromagnetic phenomena.

Importance and Impact

The discovery and understanding of lodestone's magnetic properties had far-reaching implications in both science and society. Its natural magnetic force was not only a curiosity but also a gateway to understanding the **Earth's magnetic field** and, eventually, the broader concept of electromagnetism. By investigating lodestone, early scientists like William Gilbert revealed that magnetism was not a supernatural force but a natural property, challenging superstitions and inspiring a new wave of scientific inquiry. This shift in understanding magnetism paved the way for the development of physics as a more rigorous field, bridging natural philosophy with systematic scientific methodology.

The practical applications of lodestone, particularly in navigation, had a profound impact on human history. The magnetic compass, derived from lodestone's properties, revolutionized maritime exploration, allowing sailors to venture further than ever before. This technological advancement enabled **global trade**, **exploration**, and **cultural exchange**, laying the foundation for the interconnected world that would follow. The compass was indispensable during the Age of Exploration, guiding explorers such as **Christopher Columbus** and **Vasco da Gama** across uncharted seas, reshaping economies, and establishing new trade routes that altered the course of history.

In the realm of scientific development, lodestone served as an early foundation for the study of magnetism and electricity, fields that would eventually merge in the 19th century into the discipline of **electromagnetism**. Gilbert's theories inspired later scientists, including **Hans Christian Ørsted** and **Michael Faraday**, to investigate the relationship between electricity and magnetism, ultimately leading to the discovery of **electromagnetic induction** and the creation of the first electric generators. Thus, the study of lodestone indirectly contributed to advancements that would power the modern world, from electricity generation to telecommunications.

The story of lodestone also serves as a reminder of the importance of **natural phenomena** in inspiring scientific discovery. Long before there were theories or equations to explain magnetism, there was lodestone—a simple rock that prompted questions, speculation, and experimentation. This natural curiosity-driven journey from lodestone to electromagnetism highlights the incremental nature of scientific progress, where even the most basic observations can lead to groundbreaking discoveries. The journey from lodestone to the magnetic compass to a scientific understanding of the Earth's magnetic field exemplifies how ancient observations can culminate in transformative technologies and scientific theories.

Chapter Summary

The discovery of lodestone's magnetic properties marks a pivotal moment in the history of science and technology. What began as a curiosity among ancient civilizations evolved into a powerful tool for navigation, enabling the compass and ushering in an era of exploration that would change the world. Lodestone's magnetism led to foundational insights into the Earth's magnetic field, inspiring early studies of magnetism and laying the groundwork for the field of electromagnetism. Figures like William Gilbert expanded on these properties, moving magnetism from mysticism to science. Lodestone thus stands as an emblem of how natural phenomena, carefully observed, can inspire questions and spark a chain of discoveries with lasting impact.

Next Chapter

With magnetism moving from myth to science, other natural mysteries began to draw scientific scrutiny. Like magnetism, the **idea of spontaneous generation**—the notion that life could arise from non-living matter—had captured imaginations for centuries. Yet, much like early ideas about magnetism, this theory would soon be overturned. In the next chapter, we'll enter the lab of **Louis Pasteur**, who, with a simple flask experiment, would refute spontaneous generation and lay the foundations for **modern microbiology** and **germ theory**, marking a turning point in the understanding of life itself.

Spontaneous Generation's Fall
A Simple Flask and the Birth of Microbiology

For centuries, the idea of **spontaneous generation**—the belief that life could arise from non-living matter—dominated scientific thinking. Ancient philosophers, including **Aristotle**, theorized that certain life forms, such as flies or maggots, appeared naturally from decaying organic material. This theory, while based on observation, remained unchallenged for generations, especially since the tools needed to study microscopic life were not yet available. The idea of life spontaneously appearing from non-life fit comfortably into the worldview of the time, bridging the mysterious gap between the living and non-living.

By the 19th century, however, advances in **microscopy** began to cast doubt on spontaneous generation. Scientists now observed microorganisms invisible to the naked eye, which led to new questions about the true origins of life. **Louis Pasteur**, a chemist and microbiologist from France, became one of the foremost challengers of spontaneous generation. Known for his work on fermentation and silkworm diseases, Pasteur was deeply interested in the origins and behavior of microbial life. Around the 1850s, Pasteur set out to demonstrate that the theory of spontaneous generation was incorrect, suspecting that life always originated from other life forms.

To test his hypothesis, Pasteur designed an experiment using a **swan-necked flask**. He filled the flask with nutrient-rich broth, boiled it to sterilize its contents, and allowed it to cool. The unique shape of the swan neck prevented airborne microorganisms from reaching the broth while still allowing air to enter. According to spontaneous generation theory, life should emerge within the broth on its own; however, Pasteur's flask remained free of microbial growth as long as the neck stayed intact. When he tipped the flask or broke the neck, allowing microorganisms in the air to enter the broth, life quickly appeared, proving that microorganisms originated from external sources. With this experiment, Pasteur definitively demonstrated that life did not arise spontaneously from non-life.

Importance and Impact

Pasteur's experiment with the swan-necked flask marked a **turning point** in biology and medicine. By disproving spontaneous generation, Pasteur introduced the concept of **biogenesis**—that life arises only from existing life. This principle became foundational in microbiology, reshaping the study of life and influencing fields such as **medicine**, **biology**, and **food safety**. Pasteur's findings established that microbes were responsible for processes like fermentation and decay, and, more significantly, that they could affect human health. The idea that disease could be caused by microscopic life was revolutionary, setting the stage for the **germ theory of disease**.

The implications of Pasteur's discovery were far-reaching. His work led directly to the development of **aseptic techniques** in medicine and laboratory work. By showing that microorganisms could be prevented from contaminating sterile environments, Pasteur influenced practices that would reduce infection rates in surgeries and improve hygiene standards in hospitals. These techniques, adopted by pioneers like **Joseph Lister** in antiseptic surgery, saved countless lives by reducing the spread of infectious diseases, transforming medical practice from the mid-19th century onward.

Pasteur's findings also had a profound impact on industries beyond medicine. His research in microbial life extended to **food preservation**, leading to the creation of the **pasteurization process**. In pasteurization, liquids such as milk and wine are heated to kill harmful microbes, extending their shelf life and reducing the risk of foodborne illnesses. Pasteur's innovation in food safety was essential for urban populations, allowing perishable foods to be transported and stored more safely. Today, pasteurization remains a critical process in the food and beverage industries worldwide, a direct legacy of Pasteur's early experiments.

Pasteur's success inspired other scientists to delve deeper into the microbial world. Figures like **Robert Koch** built upon Pasteur's discoveries, identifying specific microbes responsible for diseases like tuberculosis and cholera. This marked the beginning of **bacteriology** and the targeted study of pathogens, which ultimately led to the development of vaccines and antibiotics. Pasteur's work also fueled public health reforms, encouraging sanitation measures that reduced the spread of infectious diseases. His contributions to microbiology and public health underscored the idea that scientific discoveries could directly improve the quality of life, driving forward the field of modern medicine.

Chapter Summary

Louis Pasteur's work with the swan-necked flask experiment conclusively disproved the theory of spontaneous generation and established that life originates only from pre-existing life. His findings laid the foundation for germ theory, reshaping medicine, hygiene, and food preservation. Pasteur's research introduced aseptic techniques that transformed surgical practices, created the process of pasteurization for safe food storage, and inspired a generation of scientists to study pathogens, paving the way for advancements in microbiology, immunology, and public health. Pasteur's legacy continues to impact medicine and everyday life, illustrating the power of science to challenge long-held beliefs and drive human progress.

Next Chapter

With the mystery of spontaneous generation resolved, scientists began exploring other invisible forces in nature. Like microorganisms, **sound waves** were an unseen yet powerful phenomenon, and capturing sound in a visible form became the next challenge for inventors. In the following chapter, we'll meet **Édouard-Léon Scott de Martinville**, a French printer who pioneered sound visualization with his invention of the **phonautograph**. This groundbreaking device marked the first attempt to record sound waves, setting the stage for future advances in audio technology.

007 Seeing Sound
Visualizing Sound Waves for the First Time

In the mid-19th century, scientific curiosity about sound was beginning to grow, although its properties remained a mystery. While sound was clearly perceptible to the ear, its physical nature and behavior were difficult to study since sound waves were invisible. Amid this scientific curiosity, **Édouard-Léon Scott de Martinville**, a French printer with a fascination for acoustics, wanted to find a way to capture sound in a form that could be observed and analyzed. Scott's passion was particularly focused on the **human voice**, and he envisioned a method that would allow people to visualize vocal sounds and study their properties. However, Scott was not a trained scientist or engineer; instead, he was a self-taught inventor whose novel ideas would lead him to a groundbreaking, albeit underappreciated, achievement.

In 1857, Scott invented the **phonautograph**—the world's first device capable of recording sound visually. The phonautograph was a relatively simple yet ingenious device: it consisted of a **horn-shaped funnel** that funneled sound waves onto a thin membrane. Attached to the membrane was a lightweight stylus, which vibrated in response to the sound waves and traced these vibrations as **wavy lines** onto a smoke-blackened glass plate or a paper cylinder. Although the phonautograph could not play back the recorded sounds, it transformed sound waves into a visible form, providing scientists with a way to study the shape and amplitude of sound for the first time.

Scott's phonautograph was inspired by his love of language and speech. He hoped that his device would help linguists and phoneticians analyze speech patterns and vocal sounds, contributing to the scientific study of language. While the phonautograph attracted some interest among scientists, it was largely seen as a curiosity rather than a practical invention. Scott's creation ultimately predated **Thomas Edison's phonograph**—the first device to play back recorded sounds—by two decades. Because it could not reproduce sound, the phonautograph did not gain widespread recognition or commercial success in Scott's lifetime. However, in recent years, some of Scott's original recordings, called phonautograms, were rediscovered and successfully converted into audible sound using modern technology. In these recordings, Scott's own voice can be heard singing "Au Clair de la Lune," giving us a rare auditory glimpse into his 19th-century world.

Importance and Impact

Although the phonautograph itself did not evolve into a widely used technology, its invention represented a pivotal step in the study of **acoustics and audio recording**. By creating the first visual representation of sound waves, Scott made it possible for scientists to examine the physical characteristics of sound in ways that had never before been possible. The wavy lines recorded by the phonautograph allowed researchers to analyze aspects of sound such as **frequency**, **amplitude**, and **waveform**, providing a foundation for the scientific understanding of sound. Scott's device essentially marked the beginning of **audio visualization**, a field that would later become crucial in both scientific research and audio engineering.

The phonautograph directly influenced subsequent developments in sound recording technology. It provided a conceptual foundation for future inventors, including Thomas Edison, who would later create the phonograph—a device that could both record and play back sound. While Scott's phonautograph could not accomplish this, it demonstrated that sound could indeed be captured and studied, inspiring Edison and others to pursue audio technology further. The shift from mere visualization to **audible playback** was transformative, making it possible for sound to be preserved, shared, and replayed, a development that revolutionized communication, entertainment, and language documentation.

The impact of Scott's phonautograph also extended to **linguistics and speech analysis**. By producing visual traces of sound, the device allowed scientists to begin studying speech patterns with a new level of precision, laying groundwork that would influence later research in fields like **speech therapy**, **phonetics**, and **forensic linguistics**. Although Scott's device was initially intended for linguistic analysis, its legacy would ultimately spread across disciplines, inspiring fields such as sound engineering, music production, and digital signal processing. The phonautograph reminds us that groundbreaking inventions often emerge from simple observations and basic curiosity.

In modern times, the rediscovery and playback of Scott's phonautograms have attracted considerable interest. These early recordings not only highlight Scott's pioneering role in sound recording but also capture a unique piece of historical heritage. Listening to these recordings today offers a rare opportunity to hear a human voice from the 1850s, linking past and present through technology. Scott's achievement, once forgotten, has thus been celebrated as a milestone in the history of sound science, and his invention has been recognized as a key precursor to modern audio recording. The preservation of his work underscores the importance of archiving scientific artifacts, as advances in technology can breathe new life into past discoveries.

Chapter Summary

Édouard-Léon Scott de Martinville's phonautograph marked the first successful attempt to record sound visually, using wavy lines to represent sound waves on soot-blackened surfaces. Although the phonautograph could not play back sounds, it provided scientists with a new way to study the properties of sound waves, laying the foundation for the fields of acoustics and sound recording. Scott's pioneering work inspired future inventors, most notably Thomas Edison, who would later develop the phonograph, the first device capable of replaying recorded sounds. Rediscovered in recent years, Scott's phonautograms have been transformed into audible recordings, allowing listeners today to hear his voice—a remarkable bridge to the early days of audio science. His phonautograph stands as a testament to the power of curiosity and the enduring impact of even the most modest scientific innovations.

Next Chapter

With Scott's phonautograph capturing the invisible waves of sound, researchers became increasingly interested in understanding other invisible forces at work in the natural world. In the following chapter, we'll delve into the story of **Joseph Priestley** and his accidental discovery of **oxygen**, an element that would forever transform chemistry and reveal the hidden properties of air, reshaping our understanding of combustion, respiration, and life itself.

Serendipitous Gas

An Exploding Balloon and the Discovery of Oxygen

The late 18th century was a period of profound curiosity and experimentation among scientists. Chemistry, in particular, was on the cusp of major transformations. Scientists had long recognized various gases like "fixed air" (carbon dioxide) and "inflammable air" (hydrogen), but the very nature of air itself remained unclear. **Joseph Priestley**, an English theologian, philosopher, and amateur scientist, was one of those deeply interested in these mysterious "airs." Priestley's approach to science was unconventional, as he didn't have formal training, yet his experiments revealed remarkable findings, often by accident. Fascinated by the reactions between different substances, Priestley conducted numerous tests in his modest home laboratory, creating a groundbreaking revelation through a mixture of curiosity, intuition, and chance.

In 1774, Priestley directed his experiments towards understanding how various materials reacted with sunlight and heat. He used a **burning glass** to focus sunlight onto mercuric oxide, a substance he placed inside a glass container. As the sunlight heated the mercuric oxide, it released a gas that Priestley could observe accumulating in the container. Curious about the properties of this gas, Priestley placed a candle in the container and was surprised to see it burn much brighter than in ordinary air. Intrigued, he placed a small mouse inside the container, noting that the animal was able to live longer than in regular air. This led Priestley to conclude that the gas was unusually supportive of **combustion and respiration**, suggesting that it was a vital component of air.

Priestley named this newly discovered gas **"dephlogisticated air"**, according to the phlogiston theory popular at the time, which held that combustion involved a substance called phlogiston. In reality, Priestley had discovered **oxygen**, although the true significance of this discovery would take time to be fully understood. Priestley's experiment soon attracted the attention of other prominent chemists, including **Antoine Lavoisier** in France. Lavoisier would later disprove the phlogiston theory, showing that Priestley's "dephlogisticated air" was a new element—oxygen—and was essential to combustion. Priestley's accidental discovery of oxygen would go on to fundamentally alter the field of chemistry and inspire a new understanding of **elements** and **reactions**.

Importance and Impact

Priestley's accidental discovery of oxygen was one of the most important breakthroughs in the history of **chemistry**. Although Priestley himself did not fully grasp the implications of his discovery, it opened the door for the **chemical revolution** that would soon unfold. When Lavoisier demonstrated that combustion and respiration were processes involving this new gas, it challenged and ultimately overturned the phlogiston theory that had dominated scientific thought for decades. This marked the beginning of **modern chemistry**, where substances were understood in terms of elements and compounds, rather than mystical substances like phlogiston. The identification of oxygen as an essential element laid the groundwork for the periodic table and enabled scientists to classify and understand matter in ways that had never been possible before.

Priestley's work also had profound implications for **biology** and **medicine**. By recognizing that oxygen was crucial to respiration, scientists began to investigate its role in sustaining life, setting the stage for the later discovery of cellular respiration and the mechanisms by which oxygen is transported through the bloodstream. This understanding has since become essential to medicine, from resuscitation techniques to modern therapies for respiratory illnesses. The concept that a particular element was central to life itself transformed biology and led to life-saving advancements in medical treatments and emergency care. Oxygen's importance in combustion also had practical applications, enabling advancements in metallurgy and manufacturing, which required controlled and intensified flames.

Beyond chemistry and biology, Priestley's discovery illustrated the role of **serendipity** in scientific advancement. Though he did not set out to discover oxygen, his curiosity and willingness to experiment led him to findings that he could not have anticipated. Priestley's unexpected results underscored the value of open-minded investigation and encouraged other scientists to conduct their own experiments, even if the outcomes were uncertain. His work reminded the scientific community of the importance of observation, documentation, and the testing of assumptions. Priestley's discovery of oxygen is a prime example of how chance, combined with curiosity and persistence, can produce transformative knowledge.

The international impact of Priestley's work also set a precedent for **collaboration** and **exchange** among scientists. His findings crossed borders, reaching Lavoisier in France and sparking a scientific dialogue that led to major shifts in understanding. The discovery of oxygen became a focal point in the history of science, illustrating the way breakthroughs often involve contributions from multiple thinkers across different cultures. The collaborative interpretation of Priestley's findings by Lavoisier and others created a foundation for the global scientific community, where knowledge is shared, challenged, and refined.

Chapter Summary

Joseph Priestley's serendipitous discovery of oxygen marked a transformative moment in science. Through his simple experiment with a burning glass and mercuric oxide, Priestley uncovered a gas that was crucial to combustion and respiration. Though he did not fully realize its significance, his findings inspired Antoine Lavoisier to define oxygen as a fundamental element, leading to the abandonment of the phlogiston theory and the birth of modern chemistry. Oxygen's discovery would become essential to medicine, biology, and industry, establishing Priestley's work as a cornerstone of scientific advancement. His story highlights the value of curiosity-driven experimentation and the role of chance in the discovery process, reinforcing the idea that science often progresses in unexpected ways.

Next Chapter

With the identification of oxygen, science had shown that unseen forces—like gases—could be central to life and fundamental processes. Another unseen force, **magnetism**, would continue to fascinate researchers, who sought to understand how magnetic fields shaped Earth's behavior. In the next chapter, we'll delve into the **mysterious magnetic poles** and early explorations into geomagnetism, as scientists began to unravel the secrets of Earth's magnetic properties and their implications for navigation and planetary science.

Mysterious Magnetic Poles

The Compass that Shifted Understanding of Magnetism

The use of magnetism for navigation dates back many centuries, with early records suggesting that ancient Chinese civilizations were among the first to recognize and utilize this strange force. As early as the 11th century, Chinese mariners were using **lodestone** (a naturally occurring magnetic rock) as a primitive compass to aid in their voyages. This development marked a revolutionary step in navigation, providing a tool that could guide travelers even in poor visibility, transforming exploration and trade. By the 12th century, knowledge of the magnetic compass had spread across the Islamic world and into Europe, where it would become essential for seafaring journeys.

While the magnetic compass was invaluable to explorers, it was also a profound mystery. It was unclear why lodestone, or any magnetic material, had the power to indicate direction, let alone why it consistently pointed north and south. Scholars and philosophers grappled with this question, intrigued by the existence of **magnetic poles** that seemed to exert an invisible influence over the compass. The discovery of these poles—mysterious yet essential for navigation—challenged the scientific understanding of natural forces, prompting inquiries that would lead to significant advancements in both **physics** and **earth science**.

In Europe, during the 13th and 14th centuries, scholars such as **Petrus Peregrinus de Maricourt** began investigating the nature of magnetic materials and their behavior. Peregrinus documented how lodestone could be used to construct a functional compass, and he observed that a magnet's influence seemed strongest at two opposite ends—what we now call magnetic poles. In his seminal 1269 treatise Epistola de Magnete, Peregrinus described magnetic polarity, marking one of the earliest scientific accounts of magnetism. His observations paved the way for future inquiries into magnetism and the properties of the Earth itself.

However, it would take several more centuries to gain a deeper understanding of magnetic poles and the Earth's magnetic field. This slow but steady accumulation of knowledge would eventually lead scientists to realize that **Earth itself behaves like a giant magnet** with poles that guide compasses. This notion laid the groundwork for the scientific discipline of geomagnetism, which would seek to explain the nature of the magnetic field and its fluctuations over time. The quest to understand magnetic poles and the compass's north-south alignment would ultimately lead to significant discoveries about the Earth's structure and magnetic behavior.

Importance and Impact

The recognition and understanding of **magnetic poles** were instrumental in advancing navigation and exploration. The magnetic compass allowed sailors to traverse open seas and unknown territories with confidence, sparking an era of global exploration that reshaped history. By providing explorers with a reliable means to determine direction, the compass enabled journeys to previously unreachable destinations, paving the way for **cultural exchange**, **trade expansion**, and **geographic discovery**. This tool contributed to the success of explorers like **Marco Polo** and later, **Christopher Columbus**, who relied on compasses to venture into unfamiliar waters.

The compass also inspired a deeper scientific curiosity about the **forces that govern the Earth**. Scholars began to investigate why the compass consistently aligned with the north-south direction, leading to early studies of magnetism and laying the foundation for **geomagnetic science**. The notion of magnetic poles challenged natural philosophers to seek explanations beyond the visible world, fostering a spirit of inquiry that would influence scientific thought for centuries. The study of magnetism initiated an intellectual journey that would later inspire scientists like **William Gilbert** in the 16th century, whose work De Magnete would mark a major milestone in the scientific understanding of magnetism.

In addition to its practical applications, the concept of magnetic polarity contributed to early studies in **physics** and **astronomy**. By examining magnetic behavior, scientists began to contemplate the presence of unseen forces at work, leading to groundbreaking ideas that would culminate in modern electromagnetic theory. The notion of an invisible force guiding the compass needle served as a stepping stone to the later discovery of electric fields, magnetic forces, and ultimately, the development of electromagnetism. The work of early scholars on magnetic poles helped to inspire a scientific framework for studying **invisible forces** that would underpin many technological advancements in the future.

Understanding magnetic poles also had significant implications for the study of Earth's structure and dynamics. The magnetic field, with its fluctuating poles, became an area of intense research, ultimately revealing that Earth's core generates its magnetic properties. The discovery of pole reversals—periodic shifts in Earth's magnetic poles—would lead scientists to profound insights into **plate tectonics** and **geologic history**, revealing how the Earth has changed over millions of years. This link between magnetic behavior and the planet's structure opened up a new field of study that continues to impact **earth sciences** and **geology** today.

Chapter Summary

The identification and study of magnetic poles revolutionized navigation and initiated scientific inquiries that would shape fields ranging from physics to earth science. The magnetic compass, first developed with lodestone in ancient China and later refined in Europe, became an indispensable tool for explorers, sparking an era of global exploration and cultural exchange. The consistent alignment of the compass to the north-south direction led scholars to investigate magnetic forces, culminating in the recognition that Earth itself acts as a giant magnet with its own magnetic field. This understanding laid the foundation for geomagnetism and inspired future studies into electromagnetism and geologic processes. The discovery of magnetic poles marks a critical point in the human quest to understand the natural forces that shape our world.

Next Chapter

As scientists uncovered the mysteries of magnetism, the pursuit of understanding invisible forces expanded to include another enigmatic phenomenon: **electricity**. Just as magnetic poles had puzzled early researchers, so too did the nature of electric currents and conductivity. In the next chapter, we will explore **Benjamin Franklin's famous kite experiment** and the early discoveries that helped to define electricity, leading to a better understanding of the forces that power our world.

010 Taming Electricity
A Key, a Kite, and the Discovery of Conductivity

In the 18th century, **electricity** was a mysterious phenomenon, observed in lightning storms and occasionally in laboratory experiments, yet largely misunderstood. Scholars and scientists were aware of static electricity, but its nature and potential uses remained elusive. This began to change in the mid-1700s, thanks to the curiosity and experimentation of **Benjamin Franklin**, an American polymath who was fascinated by the possibilities of electricity. Franklin had already contributed to society through his work in printing, science, and public affairs, yet he was determined to understand electricity more deeply. His approach was unconventional and involved methods that others might have considered dangerous or even foolish.

Franklin's most famous experiment took place during a thunderstorm in 1752, when he flew a **kite** with a **metal key** attached to its string. This experiment, though risky, was designed to test his hypothesis that **lightning was a form of electricity**. Franklin suspected that storm clouds carried an electrical charge that could be harnessed, similar to the way static electricity behaves in controlled settings. As the storm intensified, Franklin noticed that the loose fibers on the string began to stand on end, indicating the presence of an electric field. Eventually, sparks jumped from the key to his knuckle, confirming that the kite had conducted **electricity** from the storm clouds above. Through this daring experiment, Franklin demonstrated that lightning was indeed an electrical phenomenon.

While Franklin's kite experiment is well-known, its implications were revolutionary for the study of electricity. By proving that lightning was a natural form of electricity, he introduced the concept of **conductivity**—the ability of certain materials to carry an electrical charge. This understanding of conductivity opened up a world of possibilities for controlling and utilizing electricity. Following this breakthrough, Franklin went on to invent the **lightning rod**, a device designed to protect buildings by safely conducting electrical charges from lightning strikes into the ground. This practical application of his discoveries marked one of the first times that scientific knowledge about electricity was translated into a device with real-world benefits, establishing Franklin not only as an experimenter but also as an inventor.

Importance and Impact

Franklin's kite experiment had a profound impact on both science and society. By establishing a connection between lightning and electricity, he provided the first significant evidence that electricity was not only a laboratory curiosity but also a natural force that could be studied and controlled. This breakthrough helped dispel mystical and superstitious beliefs about lightning, transforming it from a frightening, unpredictable phenomenon into a subject of scientific inquiry. Franklin's work inspired further research on electricity, particularly the study of **conductive materials** and electrical charges, which would later prove essential to the development of modern electrical engineering.

One of the most immediate outcomes of Franklin's discoveries was the invention of the **lightning rod**. Prior to this invention, lightning strikes were a frequent cause of fires and damage to buildings, particularly in densely populated urban areas. Franklin's lightning rod provided a simple but effective way to protect structures by channeling the electrical charge safely into the ground. This invention rapidly gained popularity across Europe and North America, reducing the risk of fires from lightning and saving lives and property. The lightning rod became an enduring symbol of Franklin's contribution to science and showcased how theoretical knowledge could be applied to solve practical problems.

Franklin's demonstration of conductivity also laid foundational principles for **electricity research** that would later lead to significant technological advances. His insights on the behavior of electrical charges and conductive materials influenced future scientists, including **Alessandro Volta**, **Michael Faraday**, and **James Clerk Maxwell**. Volta's development of the electric battery, Faraday's discovery of electromagnetic induction, and Maxwell's equations all drew on principles of electricity and conductivity that Franklin had helped establish. These advancements eventually enabled the invention of the electric motor, the telegraph, and later, the entire field of electrical engineering, fundamentally reshaping society and industry.

Franklin's work highlighted the importance of **curiosity-driven experimentation** and the willingness to take risks in the pursuit of knowledge. His kite experiment was both dangerous and unconventional, yet it yielded insights that would alter humanity's understanding of electricity. Franklin's approach exemplified the Enlightenment spirit, encouraging others to question established beliefs and pursue scientific exploration. By demonstrating that natural forces like electricity could be understood and harnessed, Franklin helped lay the groundwork for the future of scientific discovery. His contributions served as an inspiration to inventors and scientists for generations to come, proving that experimentation, even with humble materials, could lead to groundbreaking insights.

Chapter Summary

Benjamin Franklin's famous kite experiment marked a turning point in the study of electricity by proving that lightning was a natural form of electricity. His daring experiment introduced the concept of conductivity and established that electrical charges could be harnessed and controlled. This breakthrough led to the invention of the lightning rod, a practical application that saved countless lives and property. Franklin's insights on conductivity laid the groundwork for future discoveries in electrical science, influencing figures like Volta, Faraday, and Maxwell, whose work would shape the future of technology. Franklin's approach to science, characterized by curiosity and innovation, remains a timeless example of how observation and experimentation can lead to discoveries with far-reaching impact.

Next Chapter

With Franklin's work on conductivity sparking a new era of experimentation in electricity, scientists began to explore the effects of electrical currents on living organisms. In the next chapter, we'll examine the experiments of **Luigi Galvani**, who discovered a curious connection between electricity and muscle movement through his studies with frog legs. This unexpected link between electricity and biology would lead to the birth of **bioelectricity** and provide the foundation for advances in neurophysiology and medical technology.

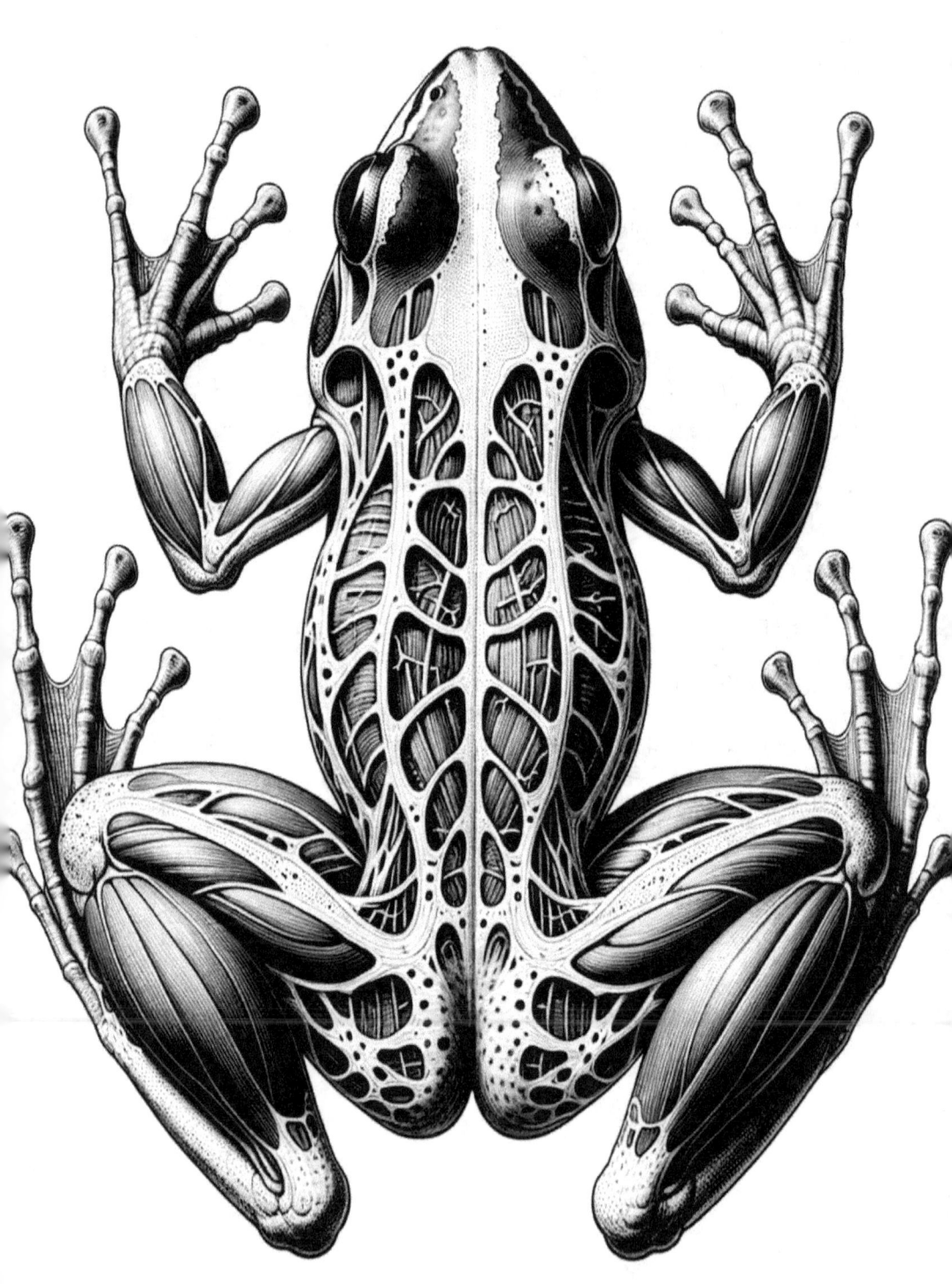

Galvani's Frogs
Muscle Twitches and the Origins of Bioelectricity

In the late 18th century, the nature of **electricity** and its possible connections to life were subjects of intense curiosity and debate. Scientists had observed the phenomenon of static electricity and experimented with rudimentary electrical devices, yet the mechanisms by which electricity functioned within living organisms were not well understood. Among those intrigued by the mysteries of electricity was **Luigi Galvani**, an Italian physician and scientist. Galvani's interest in the physiological effects of electricity led him to conduct experiments on **animal tissues**, especially **frog legs**. He believed that studying the way muscles responded to electricity might reveal some hidden aspect of life itself.

In 1780, while conducting experiments at the University of Bologna, Galvani observed something remarkable. He discovered that when he applied **metallic probes** to a dissected frog's leg, the muscle twitched as though alive. This muscle contraction seemed to occur spontaneously whenever he introduced a metallic contact, even in the absence of an external electrical source. This led Galvani to theorize that there was an intrinsic "animal electricity" within the frog, a natural force within living tissues that responded to external metal stimuli. He believed he had discovered a new force inherent to living organisms, separate from the "artificial" electricity generated by machines like the early batteries of the time.

Galvani's observations became the foundation of his concept of **bioelectricity**—the idea that animals produce their own natural electricity. He published his findings in 1791, proposing that this "animal electricity" was generated within the body and was the force behind muscle movement. Galvani's research garnered considerable attention, both from admirers who saw potential for new discoveries and skeptics who believed that his observations could be attributed to external electrical sources. One of the most vocal of these skeptics was **Alessandro Volta**, another Italian scientist who would soon become Galvani's scientific rival.

Importance and Impact

Galvani's experiments on frogs' legs marked a turning point in **physiology** and the understanding of **electricity in biology**. Although his theory of "animal electricity" would eventually be revised, Galvani's work introduced the idea that electrical forces could be intrinsic to living organisms, a groundbreaking concept that laid the groundwork for bioelectricity as a field. His observations of muscle contractions under metallic influence hinted at the **electrical nature of nerves and muscles**, which would later prove foundational in understanding how the body transmits signals and controls movement.

The public was captivated by Galvani's findings, and the sight of twitching frog legs became a staple of scientific demonstrations across Europe, blending science with spectacle. While many were fascinated by the idea of "animal electricity," Galvani's work inspired further inquiry into the nature of electricity in both living and non-living systems. His findings contributed to the perception that electricity might have powerful applications in medicine, potentially aiding in the treatment of paralysis and other neurological disorders, which scientists and doctors would later pursue through the development of **electrical stimulation therapies**.

The debate between Galvani and Volta over the nature of electricity would become legendary. Volta, skeptical of Galvani's conclusions, argued that the twitching muscles were due to **external electricity**, generated by the metal contacts rather than an intrinsic force within the frog's leg. To prove his theory, Volta developed the first **voltaic pile**—a primitive battery composed of alternating layers of zinc and copper, with salt-soaked paper acting as an electrolyte. This invention not only confirmed Volta's hypothesis but also became the world's first continuous source of electricity, revolutionizing the study of **electromagnetism** and launching the field of **electrochemistry**.

Despite Volta's refutation of Galvani's theory of "animal electricity," Galvani's work spurred research into **neurophysiology** and the nature of nervous system impulses. Scientists eventually demonstrated that **nerve cells** use electrical signals to communicate, validating part of Galvani's hypothesis about bioelectric forces within living tissues. Galvani's legacy continued through this scientific advancement, influencing later discoveries about the **nervous system**, **muscle function**, and **electrical conduction** in cells. His findings also contributed to the development of medical devices like **pacemakers** and **defibrillators**, which rely on electrical impulses to regulate heartbeats, a direct application of bioelectricity principles.

Galvani's experiments also inspired cultural and literary works, most famously **Mary Shelley's "Frankenstein"**. Shelley's novel, published in 1818, drew from popular demonstrations of "animal electricity" and speculation about reanimating dead tissue. Galvani's twitching frogs had fascinated Europe and prompted speculation about the powers of electricity, setting the stage for Frankenstein's fictional experiment in reanimating life. This cultural impact underscored how Galvani's work resonated far beyond the laboratory, embedding itself in the imagination of an era captivated by the mysteries of life and death.

Chapter Summary

Luigi Galvani's experiments with frog legs in the late 18th century marked the beginnings of bioelectricity, a field that would eventually reveal the electrical nature of nerve and muscle function. His observation of muscle contractions in response to metallic contact led him to propose a theory of "animal electricity," igniting both public fascination and scientific debate. Though his theory was challenged by Alessandro Volta, Galvani's work laid essential groundwork for the study of electricity within living organisms, influencing future discoveries in neurophysiology and the development of medical technologies. Galvani's legacy extended beyond science, inspiring literature and culture, as his findings prompted enduring questions about the relationship between life and electricity.

Next Chapter

As Galvani's experiments with frog legs sparked debate about electricity in living tissues, Alessandro Volta was preparing to test his own theory of electrical generation. In the next chapter, we'll follow Volta's journey as he uses the principles observed in Galvani's work to create the **first true battery**. This accidental invention of the **voltaic pile** would generate a continuous electric current, ushering in a new era in both chemistry and physics, and transforming our understanding of electricity's potential.

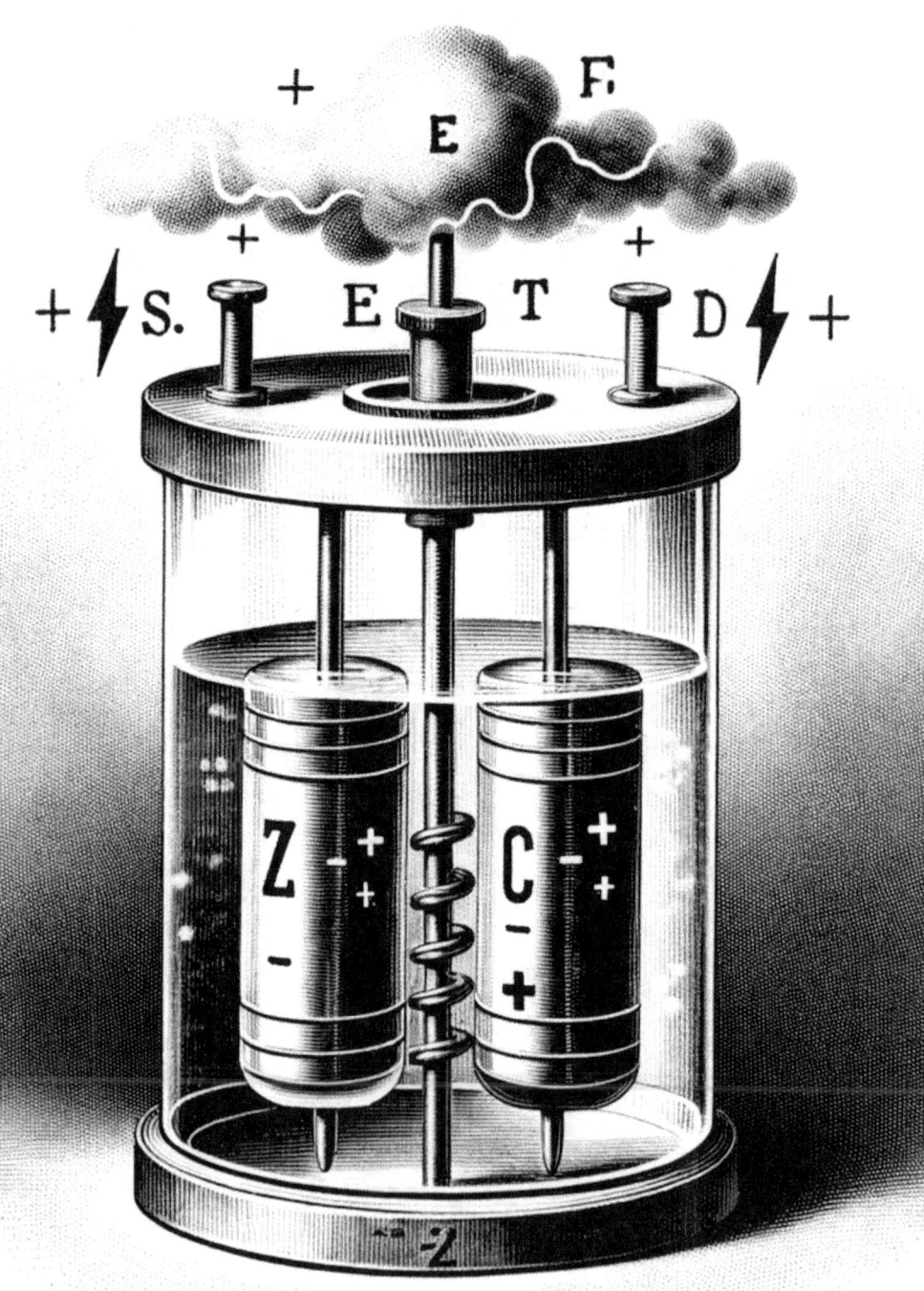

012 The Accidental Battery
From Frog Legs to the First Electric Battery

In the late 18th century, **electricity** was still a mysterious and evolving field. Luigi Galvani's recent experiments with frog legs had stirred excitement and controversy among scientists by suggesting that living tissue might generate its own form of electricity, which Galvani termed **"animal electricity"**. This idea fascinated the scientific community, but one of Galvani's peers, the Italian physicist **Alessandro Volta**, was skeptical. Volta, though intrigued by Galvani's work, believed the muscle contractions in the frogs' legs were not due to a unique animal electricity but rather the result of an external **electrical force** created by the metallic contacts themselves. Determined to prove his theory, Volta set out to investigate the cause of these mysterious muscle movements in greater depth.

Volta's work led him to explore the effects of combining different metals with a **conductive liquid**, such as saltwater, which he believed could produce an electrical current without involving any "animal" component. In 1799, through systematic experimentation, Volta discovered that by stacking alternating layers of **zinc** and **copper** separated by pieces of cloth soaked in saltwater, he could generate a steady flow of electricity. This arrangement, which he called the **voltaic pile**, became the world's first true **electric battery**. Volta had unwittingly invented a device that could produce a continuous electrical current, something previously thought impossible. This discovery would become a foundational moment in the study of **electrochemistry** and the field of electricity itself.

Volta's voltaic pile demonstrated that electricity could be generated through chemical reactions alone, independent of any biological or "animal" source, as Galvani had suggested. The pile produced electricity by the interaction of metals and electrolytes, creating a continuous flow of energy that could be harnessed and directed. Volta's work provided not only a new tool but also a new understanding of **electromotive force** and the nature of electrical currents, marking a paradigm shift from earlier theories that tied electricity exclusively to natural and isolated phenomena, like lightning or static charges.

Importance and Impact

The invention of the **voltaic pile** was monumental for both science and society, as it offered the first practical means of generating and storing **electric power**. The voltaic pile allowed scientists to conduct controlled experiments with a continuous source of electricity, something they had never had before. This breakthrough opened up new areas of study, enabling researchers to explore the properties of electric currents in-depth and setting the stage for significant advances in physics, chemistry, and engineering. Volta's battery became an essential tool for scientific research, allowing scientists to experiment with electric currents in ways that would lead to discoveries in **magnetism**, **electrolysis**, and **electrical circuits**.

Volta's invention of the battery also had a significant cultural impact. News of the voltaic pile spread quickly across Europe, inspiring both scientists and the general public with the idea that humans could produce and control electricity. This was an era fascinated by electricity's potential, and Volta's invention was seen as a step toward understanding—and perhaps even mastering—the natural world. The voltaic pile was widely celebrated and solidified Volta's reputation as a pioneering figure in electrical science. In recognition of his contributions, the unit of electric potential, the **volt**, was later named in his honor, ensuring that his legacy would live on in the fundamental language of physics.

The practical applications of the voltaic pile became apparent over the following decades. Scientists began using batteries to explore the relationship between **electricity** and **magnetism**. In 1820, **Hans Christian Ørsted** discovered that electric currents could create magnetic fields, an observation that would lead to the development of **electromagnetism**. Later, **Michael Faraday** used batteries in his experiments on **electromagnetic induction**, ultimately leading to the invention of the electric motor and generator. These inventions, made possible by the voltaic pile, laid the groundwork for the **Industrial Revolution's advances** in mechanization and power generation.

The voltaic pile also inspired the development of **electrochemistry**, as scientists began to study how electricity could induce chemical changes. Early electrolysis experiments, enabled by Volta's battery, demonstrated that electric currents could break down compounds into their constituent elements. This led to the discovery of several new elements, including **potassium** and **sodium** by **Sir Humphry Davy**. The voltaic pile thus transformed chemistry by showing that chemical bonds could be manipulated through electrical means, adding a new dimension to the understanding of atomic structure and chemical reactions.

Volta's voltaic pile remains one of the most significant breakthroughs in the history of science and technology, marking the beginning of **electric power storage** and the ability to harness electricity for practical use. His work inspired generations of scientists to continue exploring the mysteries of electricity, paving the way for modern battery technology that powers everything from mobile devices to electric vehicles. Volta's innovation not only advanced scientific understanding but also set humanity on a course toward an electrically powered future, where batteries became essential for countless applications in daily life.

Chapter Summary

Alessandro Volta's invention of the voltaic pile, the world's first electric battery, marked a turning point in the study and application of electricity. Sparked by his rivalry with Luigi Galvani, Volta set out to disprove the theory of "animal electricity" and ultimately created a device that could produce a continuous flow of electric current through chemical reactions alone. The voltaic pile revolutionized scientific research by providing a reliable power source, leading to advances in electromagnetism, electrochemistry, and power generation. Volta's work laid the foundation for countless technologies and inspired future scientists to unlock the full potential of electricity. His legacy endures in the volt, the unit of electric potential, and in the battery, a device that remains essential to modern life.

Next Chapter

With Volta's battery powering new experiments and discoveries, the potential for electricity seemed boundless. Scientists began to explore how electrical currents interacted with different materials, sparking a curiosity about **light** and **imaging**. In the next chapter, we'll follow the story of a botanist's accidental discovery of **cyanotype**, a process that would become one of the earliest methods of photography. This chance invention not only captured images but also laid the foundation for visual recording techniques, transforming the ways we document and perceive the world.

013 Cyanotype and Sunlight
A Botanist's Mistake that Birthed Photography

In the early 19th century, scientists were beginning to understand that **light could interact chemically** with certain materials. This discovery sparked curiosity about using light-sensitive compounds to capture images or record information. One of the earliest pioneers of this exploration was **Sir John Herschel**, a British astronomer, mathematician, and polymath who made significant contributions across various scientific fields. Herschel, like many scientists of his time, was fascinated by the emerging concept of "fixing" images using chemical processes. He hoped to find a practical way to make images permanent, which led him to experiment with light-sensitive compounds in the 1840s.

At the same time, **Anna Atkins**, a botanist and friend of Herschel's, was exploring new methods to document botanical specimens. Traditional hand-drawn illustrations were often labor-intensive and prone to inaccuracies. Inspired by Herschel's work, Atkins became interested in using the cyanotype process—a chemical technique Herschel had developed to create blue-toned images. Herschel had intended the cyanotype for scientific documentation, but he had no idea that it would soon become a foundational technique in photography. In the **cyanotype process**, an object is placed on paper coated with iron salts and exposed to sunlight. The light causes a reaction, leaving a permanent, intense blue image of the object's outline.

In 1843, Atkins used the cyanotype process to create a book titled Photographs of British Algae: Cyanotype Impressions, which is now considered the first published book illustrated with photographs. Her unique process enabled Atkins to produce accurate, durable images of her botanical specimens, with the vivid blue tones highlighting the fine details of leaves, stems, and petals. This **early photographic technique** allowed her to preserve her scientific observations in a visual format that was far more precise than illustrations. Though Herschel and Atkins were initially motivated by scientific goals, the cyanotype would soon be recognized as one of the first photographic techniques, laying the groundwork for the modern field of photography.

Importance and Impact

The invention of the cyanotype process and its application by Anna Atkins marked a pivotal moment in **scientific documentation** and the broader field of photography. By capturing precise, lasting images through exposure to sunlight, the cyanotype process represented a new way to record reality. Atkins's work demonstrated that photography could be used not only as an art form but also as a **scientific tool** for documenting specimens, landscapes, and data with unparalleled accuracy. Her cyanotype images provided a model for later scientific imaging techniques, from X-rays to digital scans, where capturing fine details and preserving data would become essential.

Atkins's book of cyanotype images not only made her the first person to publish a book illustrated with photographs but also helped popularize the **photographic medium** among scientists and artists. The cyanotype method, which was simple, inexpensive, and required only basic materials, allowed a wider audience to experiment with photography. In the decades that followed, photography evolved from an obscure scientific pursuit into a widely practiced form of art and personal expression, eventually revolutionizing **journalism**, **portraiture**, and **documentation**.

The cyanotype process also had a lasting impact on scientific imaging and preservation. Atkins's method of documenting plant life created a model for using photography in fields such as **botany**, **archaeology**, and **astronomy**, where detailed records were crucial. Her work directly influenced how scientists visualized and shared their findings, making data more accessible and reliable. As a result, the cyanotype process became foundational for generations of researchers who sought to capture and analyze images with greater precision. Today, cyanotype is still practiced as a form of alternative photography, celebrated for its unique color and texture, a lasting legacy of its origins in scientific experimentation.

The cyanotype also had cultural resonance beyond its scientific applications. The process became popular among artists who were drawn to its distinctive blue tones and ethereal quality, often using it to create aesthetic compositions rather than scientific records. This merging of science and art foreshadowed the ways that photography would blur boundaries between disciplines, influencing fields as diverse as **fine art**, **design**, and **advertising**. Atkins's work bridged these worlds, illustrating the potential for scientific tools to inspire creativity and shift cultural perspectives on art and documentation.

The story of the cyanotype reminds us of how **serendipitous discoveries** can have long-lasting impact. What began as a botanical documentation method became a foundation for an entirely new medium, reshaping the way humans preserve and share visual information. Herschel and Atkins likely had no idea that their work would serve as the basis for photography, yet their curiosity-driven experiments with light-sensitive materials opened doors to innovations that would transform modern culture, science, and communication.

Chapter Summary

The cyanotype process, developed by Sir John Herschel and famously applied by botanist Anna Atkins, marked a turning point in both science and photography. Initially intended as a tool for documenting botanical specimens, the cyanotype became one of the earliest photographic methods, allowing for precise, durable images through sunlight exposure. Atkins's cyanotype illustrations bridged science and art, influencing both the development of scientific imaging and the aesthetic exploration of photography. Her work demonstrated the power of photography as a tool for preservation and visualization, laying the groundwork for a medium that would revolutionize documentation and personal expression across the globe.

Next Chapter

As the cyanotype process established new possibilities for documenting the world through photography, another scientist, **Antoine Lavoisier**, was making strides in understanding the world at a molecular level. In the next chapter, we'll explore Lavoisier's work on **combustion** and the **disproof of phlogiston theory**, an advancement that would help shape modern chemistry. His experiments would provide a systematic understanding of chemical reactions, marking the birth of a new era in scientific inquiry.

014 Lavoisier's Elements
How Burning Disproved Phlogiston Theory

In the 18th century, chemistry was still rooted in mystical concepts and untested theories. One of the most popular theories of the time was the **phlogiston theory**, which proposed that a substance called phlogiston was released during combustion. According to this theory, when materials burned, they released phlogiston into the air, leaving behind a "dephlogisticated" substance. Developed in the 1600s and popularized in the 1700s, phlogiston theory attempted to explain the processes of burning, rusting, and other transformations. However, this theory had significant flaws that became evident as scientists began to measure mass changes during reactions, leading to inconsistencies in its application.

Amid this growing skepticism, a new approach to chemical reactions was emerging. **Antoine Lavoisier**, a French nobleman and scientist, was determined to bring rigor and precision to chemistry. Lavoisier believed that chemistry required a systematic approach based on **careful measurement** and **observation**. His experiments led him to suspect that something other than phlogiston was at work in combustion. Around the same time, English chemist **Joseph Priestley** had isolated a gas he called "dephlogisticated air" (what we now know as **oxygen**), which he claimed supported combustion more effectively than regular air. Intrigued by Priestley's findings, Lavoisier set out to investigate further.

Lavoisier's experiments revealed that combustion was not caused by the release of phlogiston, but rather by the **combination of oxygen with other substances**. He meticulously measured the weight of substances before and after combustion and found that the total mass remained constant, a result that contradicted phlogiston theory. This principle, which Lavoisier termed the **law of conservation of mass**, became a cornerstone of modern chemistry. Through his research, Lavoisier established that substances are composed of distinct elements, each with unique properties. His work led him to compile the first list of chemical elements, laying the foundation for the periodic table and revolutionizing chemistry.

Importance and Impact

Lavoisier's experiments and discoveries marked the beginning of **modern chemistry**. By disproving phlogiston theory, he replaced mystical concepts with an evidence-based approach that relied on precise measurements and experimentation. His introduction of the **law of conservation of mass** was transformative, establishing that in a chemical reaction, matter is neither created nor destroyed. This principle enabled scientists to predict the outcomes of reactions with greater accuracy, setting a new standard for scientific investigation. Lavoisier's approach not only elevated chemistry but also established the need for rigorous methodology across scientific disciplines.

Through his research, Lavoisier identified and named **oxygen**, along with other gases such as nitrogen, redefining the concept of an **"element"**. His classification system was one of the first attempts to organize substances based on their chemical properties, a precursor to the **periodic table**. By introducing a standardized nomenclature, Lavoisier made chemistry more accessible and uniform, allowing scientists to communicate more effectively and laying the groundwork for future advancements. His identification of oxygen's role in combustion debunked phlogiston theory and helped to explain not only burning but also respiration, as oxygen was shown to be essential for both.

Lavoisier's work had significant practical implications as well. Understanding combustion as an oxygen-based process contributed to advancements in **industry** and **metallurgy**, allowing for more efficient fuel usage and the development of new materials. This scientific progress extended into medicine as well, where knowledge of oxygen's role in respiration advanced understanding of human biology and health. Lavoisier's ideas also inspired other scientists to explore chemical processes in greater detail, leading to breakthroughs in biochemistry and pharmacology that would ultimately transform medicine.

Unfortunately, Lavoisier's contributions were not celebrated in his lifetime. As a prominent noble and tax collector, he became a target during the **French Revolution**. Despite his remarkable contributions to science, he was executed by guillotine in 1794. However, his legacy endured, and his ideas continued to shape the field of chemistry for centuries to come. Lavoisier's emphasis on careful measurement and systematic experimentation influenced generations of scientists, from his contemporaries to modern chemists.

The shift from phlogiston theory to Lavoisier's oxygen-based understanding of combustion exemplified a major turning point in scientific history: the transition from speculative, untested theories to a rigorous scientific method based on experimentation. His work demonstrated the importance of questioning established beliefs and relying on empirical evidence to build knowledge. In doing so, Lavoisier became known as the

"father of modern chemistry", a title that reflects his lasting influence on the field and his role in transforming chemistry into a true science.

Chapter Summary

Antoine Lavoisier's experiments on combustion led to the discovery of oxygen's role in chemical reactions, debunking the long-standing phlogiston theory and establishing the law of conservation of mass. His systematic approach to chemistry introduced a new standard of precision and measurement, paving the way for modern scientific methodology. Lavoisier's work not only laid the groundwork for the periodic table but also had profound implications for fields ranging from industry to medicine. Despite his untimely death during the French Revolution, Lavoisier's legacy endures, marking him as a central figure in the history of science and the father of modern chemistry.

Next Chapter

As Lavoisier's discoveries in chemistry demonstrated the transformative power of careful measurement and experimentation, scientists in other fields began to approach their studies with similar rigor. In the next chapter, we'll delve into the world of **acoustics**, where the observation of **whispering arches** would reveal surprising properties of sound. This phenomenon, observed in architecture, would become one of the earliest studies in acoustics, inspiring new ways to understand sound waves and laying the groundwork for modern audio science.

015 The Whispering Arches
The First Acoustic Phenomenon Observed in Architecture

Centuries before modern acoustics emerged as a science, people noticed unusual auditory effects in specific architectural spaces. **Medieval Europe** saw the construction of grand **cathedrals** and **monasteries** with large, vaulted spaces and carefully crafted arches. These architectural marvels often exhibited peculiar acoustic phenomena, such as the ability to carry whispers across great distances or to amplify sounds in certain areas. These effects were not fully understood, but they were fascinating, particularly in spaces where silence and secrecy were valued. One of the most notable examples of such an acoustic wonder was found in **St. Paul's Cathedral** in London, where certain arches allowed whispers to travel across the vast room, creating the famous **Whispering Gallery** effect.

The existence of these "whispering arches" intrigued builders and religious leaders alike. While it's unclear whether medieval architects intentionally designed these effects, their discovery led to the realization that **sound behaved in complex ways** in enclosed spaces. This phenomenon likely occurred due to the shape of the vaulted arches, which directed sound waves along specific paths, allowing a whisper from one side of the gallery to be clearly heard on the other side. Although this effect appeared almost magical, it was simply the result of sound waves bouncing off the curved surfaces in ways that preserved and focused their energy. People soon noticed similar effects in other sacred spaces and large structures, such as mosques and amphitheaters.

These "whispering" effects became a popular feature of grand buildings, adding an element of mystery and reverence. Patrons and architects began experimenting with different materials, angles, and designs to see how they might enhance or dampen sound in certain spaces. This accidental discovery marked the beginning of rudimentary studies in **architectural acoustics**. Although these architects lacked the scientific language to fully explain their observations, they developed an intuitive understanding of how to manipulate sound through structural design, knowledge that would later influence the fields of **acoustics** and **architecture**.

Importance and Impact

The discovery of the whispering arches phenomenon demonstrated that **sound could be manipulated** within specific structures, leading to the recognition of acoustics as an important aspect of architectural design. The realization that architectural features could amplify or carry sound in unexpected ways encouraged builders to pay closer attention to how sound behaved in enclosed spaces. Over time, this led to an evolution in design principles, as architects began considering how structure and materials could enhance or control acoustics, particularly in spaces intended for speech or music, such as churches, theaters, and lecture halls.

These early explorations into acoustics laid the groundwork for future innovations. Although architects of the medieval period did not fully understand the scientific principles behind their discoveries, their intuitive observations became the basis for the **study of acoustics** in the centuries that followed. By the time of the **Renaissance**, architects and scholars were actively experimenting with sound in architectural spaces. Structures such as the **opera houses** of Italy and the concert halls of Europe would later be meticulously designed with acoustics in mind, building on the knowledge gleaned from these early whispering galleries.

The study of architectural acoustics would eventually emerge as a scientific discipline in the 19th and 20th centuries. Figures like **Wallace Sabine**, a pioneer of architectural acoustics, developed mathematical models to understand how sound waves traveled in enclosed spaces. His work and the subsequent growth of acoustics science would lead to designs that maximized clarity, richness, and resonance, revolutionizing performance spaces and influencing modern architecture. Thus, the discovery of the whispering arches phenomenon not only enhanced cultural spaces in the medieval world but also paved the way for modern acoustics, which we now rely on in places ranging from concert halls to stadiums.

Culturally, the whispering arches phenomenon added to the mystique of sacred and communal spaces, with whispers carrying across galleries becoming a source of fascination and inspiration. This architectural feature became a symbol of **secrecy**, **reverence**, and **wonder**, enriching the spiritual experience in churches and cathedrals. Today, the whispering galleries in buildings like St. Paul's Cathedral attract visitors eager to experience this natural acoustic marvel, a testament to humanity's long-standing fascination with the mysterious qualities of sound.

The ability to shape sound within a structure marked a shift in architectural thinking. The realization that buildings could be crafted not only to house people but also to enhance or even shape human experience was profound. Architects began to recognize the potential of design to engage the senses, laying the foundation for the multisensory approach in modern architecture. The whispering arches thus became a stepping stone in humanity's journey to understand and manipulate sound, influencing how spaces are designed for centuries to come.

Chapter Summary

The whispering arches phenomenon, first observed in medieval cathedrals like St. Paul's, marked the beginnings of architectural acoustics. This accidental discovery revealed that sound could travel in surprising ways, carrying whispers across vast spaces and adding an air of mystery to sacred places. Although medieval architects lacked the scientific knowledge to fully understand these effects, their observations laid the groundwork for the later development of architectural acoustics. This phenomenon influenced the design of churches, theaters, and concert halls, ultimately shaping the study of sound in architecture. The legacy of whispering arches lives on, inspiring curiosity and wonder in those who experience it firsthand.

Next Chapter

As architects and scholars of the medieval period marveled at the mysterious behavior of sound, scientists in other fields were uncovering equally remarkable phenomena. In the next chapter, we'll delve into the surprising discovery of **penicillin**—another accidental breakthrough that would change the world. In a small, unassuming lab, **Alexander Fleming** would stumble upon a mold that killed bacteria, sparking the development of antibiotics and transforming medicine forever.

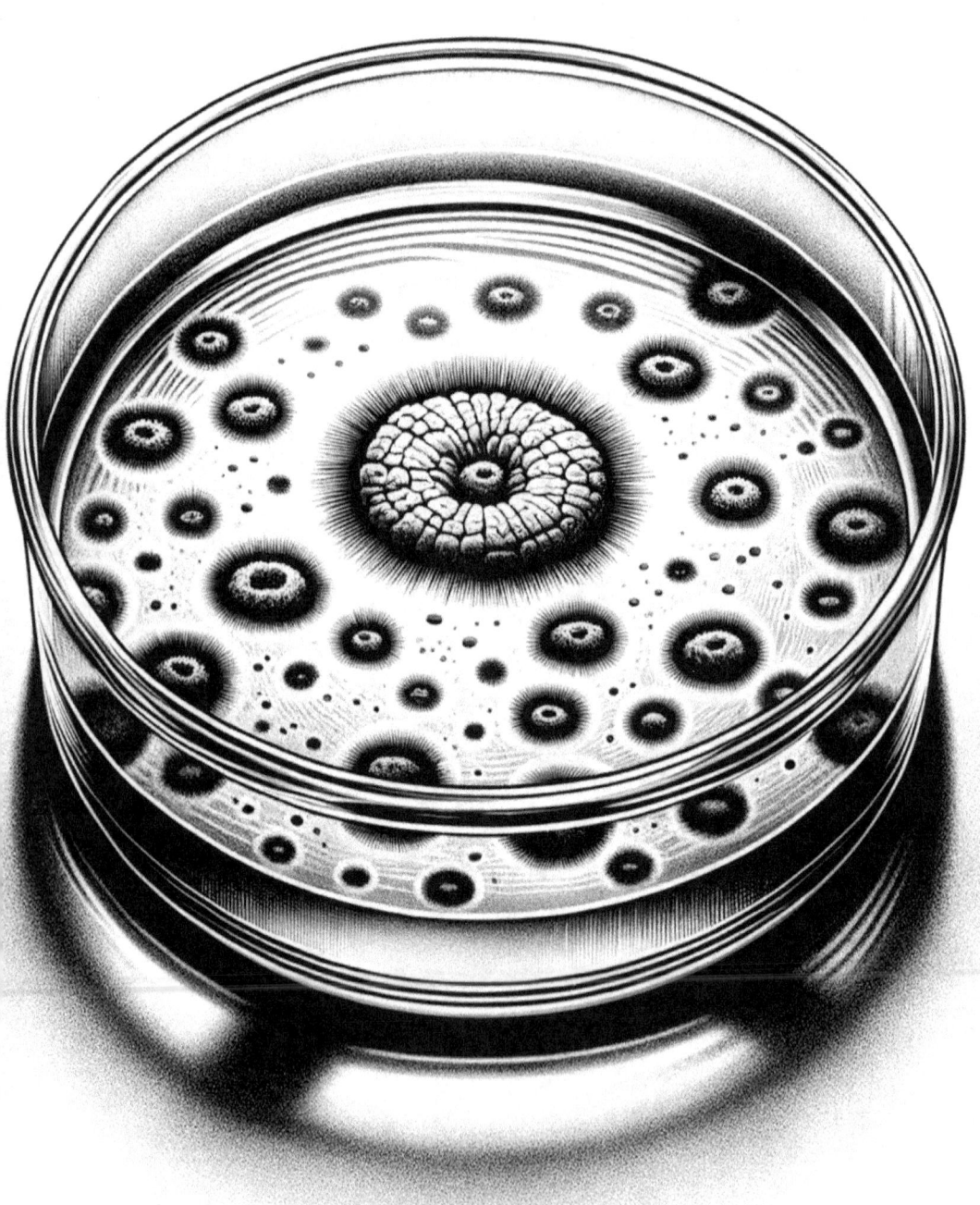

016 Penicillin's Accidental Birth
A Forgotten Petri Dish, A Bacterial Breakthrough

The dawn of the 20th century brought remarkable progress in **medical science**, yet the treatment of bacterial infections remained a significant challenge. Diseases like pneumonia, syphilis, and strep throat were often fatal, and physicians were largely powerless against them. Surgical procedures carried high risks of infection, and even minor wounds could lead to life-threatening complications. Though antiseptics like carbolic acid were used to sterilize wounds, there was no effective treatment for **systemic infections**, leaving patients vulnerable to the spread of bacteria.

In 1928, Scottish bacteriologist **Alexander Fleming** was studying the behavior of **Staphylococcus bacteria**, a common cause of infections. Working at St. Mary's Hospital in London, Fleming's experiments revolved around finding substances that could inhibit bacterial growth. However, the breakthrough for which he is now celebrated came not from meticulous experimentation but from an **unexpected accident**. After returning from a two-week vacation, Fleming noticed something unusual in one of his petri dishes, which he had left unwashed on his cluttered workbench.

The dish, which had been growing a colony of Staphylococcus bacteria, was contaminated with **mold spores**. Fleming observed that the bacteria around the mold were being destroyed, leaving a clear ring of inhibition. Intrigued, he isolated the mold and identified it as belonging to the **Penicillium genus**. Fleming hypothesized that the mold was releasing a substance that killed the bacteria. He called this substance **penicillin**. Though Fleming recognized its potential, he lacked the resources and expertise to refine penicillin into a usable medicine. For nearly a decade, his discovery remained an obscure scientific observation.

It wasn't until the early 1940s that a team of scientists at Oxford University—**Howard Florey**, **Ernst Chain**, and **Norman Heatley**—took up the challenge of mass-producing penicillin. Their work, aided by wartime funding, transformed Fleming's discovery into the first effective **antibiotic**, saving millions of lives during World War II and beyond. Fleming's accidental observation would go on to revolutionize medicine, marking the beginning of the **antibiotic era**.

Importance and Impact

Penicillin's discovery represents one of the most significant advances in the history of medicine, fundamentally altering how bacterial infections are treated. Before penicillin, infections like pneumonia and sepsis were often fatal, even with the best care available. The introduction of penicillin provided an **effective, targeted treatment**, dramatically reducing mortality rates and transforming medical practice. Diseases that were once deadly became manageable, and surgical procedures became far safer due to the reduced risk of postoperative infections.

The development of penicillin also catalyzed the growth of the **antibiotics industry**, inspiring researchers to search for other naturally occurring antimicrobial compounds. This led to the discovery of drugs like streptomycin, tetracycline, and erythromycin, which expanded the arsenal against bacterial infections. The widespread use of antibiotics reshaped public health, reducing the burden of infectious diseases and increasing life expectancy worldwide. By the mid-20th century, antibiotics had become a cornerstone of modern medicine, heralding a new era of medical treatment.

Fleming's discovery also highlighted the importance of **serendipity in scientific research**. The accidental contamination of a petri dish became a powerful example of how unplanned observations can lead to groundbreaking discoveries. However, it also underscored the need for collaboration and persistence; without the efforts of Florey, Chain, and Heatley, penicillin might never have reached its full potential as a life-saving drug. The teamwork involved in developing penicillin set a precedent for interdisciplinary collaboration in biomedical research.

The success of penicillin had profound social and economic implications as well. During World War II, the mass production of penicillin ensured that wounded soldiers could be treated effectively, reducing fatalities and infections on the battlefield. Its availability helped shape modern healthcare systems, enabling the development of **universal antibiotic programs** that dramatically improved public health outcomes. However, the widespread use of antibiotics also brought challenges, including the emergence of **antibiotic resistance**, which remains a critical issue in medicine today.

Penicillin's story is a testament to the interplay of chance, curiosity, and scientific ingenuity. Fleming's keen observation and willingness to explore an unexpected phenomenon sparked a chain of events that would change the course of medicine forever. It stands as a reminder that transformative discoveries often begin with simple, everyday observations, combined with the determination to pursue their potential significance.

Chapter Summary

The accidental discovery of penicillin by Alexander Fleming in 1928 marked the beginning of the antibiotic revolution. Observing the antibacterial properties of mold on a forgotten petri dish, Fleming identified penicillin, a substance that would later be developed into the world's first effective antibiotic. Though Fleming recognized its potential, it was the collaborative work of scientists like Howard Florey and Ernst Chain that transformed penicillin into a life-saving medicine. Penicillin revolutionized the treatment of bacterial infections, drastically reducing mortality rates and enabling safer medical practices. Its discovery reshaped public health, medicine, and society, demonstrating the power of curiosity and serendipity in driving scientific progress.

Next Chapter

While Fleming's discovery of penicillin demonstrated the potential of accidental observations in science, another unintentional breakthrough was unfolding in the field of **radiology**. In the next chapter, we'll delve into the unexpected discovery of **X-rays** by Wilhelm Röntgen in 1895, an event that would revolutionize medicine by allowing doctors to see inside the human body for the first time. This pioneering advancement opened a new window into anatomy and diagnostics, forever altering the practice of medicine.

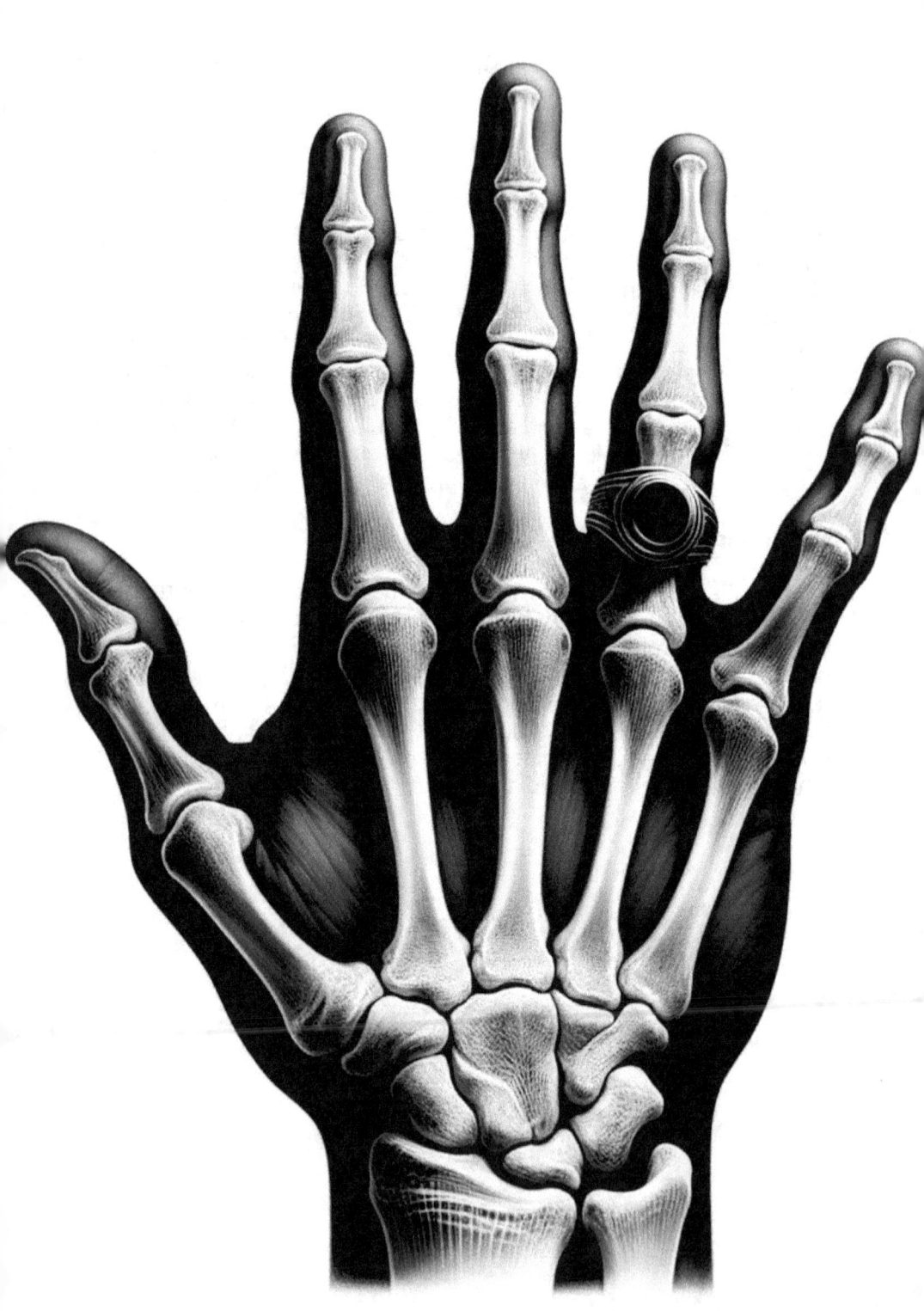

# X-Rays: Vision Beyond Sight
An Unexpected Image Reveals Inner Anatomy

In the late 19th century, the nature of light and electromagnetic waves was one of the most exciting fields in physics. Scientists had already discovered various forms of radiation, including visible light, infrared, and ultraviolet waves. Yet, the concept of waves beyond the known spectrum was still a mystery. One of the leading physicists of the time, **Wilhelm Conrad Röntgen**, was working at the University of Würzburg in Germany, investigating the behavior of **cathode rays**—streams of electrons emitted in vacuum tubes. Röntgen's goal was to understand the properties of these rays, but in 1895, his work would lead to an entirely unexpected discovery.

While experimenting with a **Crookes tube** wrapped in black cardboard to block visible light, Röntgen noticed something peculiar. A screen coated with **barium platinocyanide**, placed several feet away, began to glow faintly. This glow persisted even though the cathode rays themselves were confined within the tube. Intrigued, Röntgen deduced that an unknown type of ray, capable of passing through solid objects, was responsible for the phenomenon. He called these mysterious rays **X-rays**, using "X" to signify their unknown nature.

Röntgen soon realized the implications of his discovery when he placed various objects, including his own hand, between the tube and the fluorescent screen. To his astonishment, the screen revealed the bones of his hand, surrounded by faint shadows of flesh. This was the first recorded **X-ray image**, and it was unlike anything the scientific community had ever seen. Within weeks, Röntgen published a paper detailing his discovery, and the news spread rapidly, captivating scientists and the public alike. X-rays opened a **window into the human body**, enabling physicians to see inside without invasive procedures—a breakthrough that would revolutionize medicine.

Importance and Impact

The discovery of X-rays marked a pivotal moment in both **science** and **medicine**, transforming the way we study the natural world and diagnose illnesses. For physicians, the ability to view the internal structure of the human body without surgery was nothing short of revolutionary. Within months of Röntgen's discovery, X-ray imaging was being used in hospitals, allowing doctors to detect fractures, locate foreign objects, and diagnose diseases like tuberculosis. This new diagnostic tool significantly reduced the risks and uncertainties associated with invasive procedures, improving patient outcomes and transforming medical practice.

X-rays also had a profound impact on **physics** and the understanding of electromagnetic waves. Röntgen's discovery expanded the known electromagnetic spectrum, leading to new insights into wave behavior and energy. The study of X-rays paved the way for advancements in **quantum mechanics** and **atomic physics**, influencing future discoveries about the nature of atoms and subatomic particles. By revealing previously hidden structures, X-rays became a cornerstone for fields ranging from crystallography to materials science, enabling breakthroughs in understanding molecular structures and chemical bonds.

Despite their benefits, X-rays also brought challenges and risks. Early researchers and technicians, unaware of the dangers of prolonged exposure to X-ray radiation, suffered from burns, radiation sickness, and even cancer. These risks prompted the development of protective measures, such as **lead aprons** and shielding, as well as strict guidelines for exposure. This growing awareness of radiation hazards also spurred research into the biological effects of radiation, eventually leading to the fields of **radiation safety** and **nuclear medicine**.

The cultural impact of X-rays was equally significant. Their discovery inspired widespread fascination, with exhibitions showcasing the technology to the public. The ability to "see inside" objects seemed almost magical, blending science with wonder. X-rays became a symbol of scientific progress and the growing power of technology to reveal the unseen. This fascination extended into popular culture, where X-rays were depicted as tools of discovery, mystery, and even superhuman vision, influencing art, literature, and media.

Röntgen's discovery also earned him the **first Nobel Prize in Physics** in 1901, underscoring the importance of X-rays in advancing science and medicine. His modesty and meticulous documentation of his work became a model for scientific integrity. Although he refused to patent his discovery, believing it should benefit humanity freely, his work set the stage for a century of innovations in imaging technologies, from **computed tomography (CT)** to **magnetic resonance imaging (MRI)**. X-rays remain one of the most widely used diagnostic tools in modern medicine, a testament to the enduring legacy of Röntgen's accidental yet transformative discovery.

Chapter Summary

Wilhelm Röntgen's accidental discovery of X-rays in 1895 opened a new dimension in science and medicine, allowing us to see the invisible. By enabling the visualization of internal structures without invasive procedures, X-rays revolutionized medical diagnosis and treatment, saving countless lives. Their impact extended beyond medicine, shaping our understanding of electromagnetic waves, atomic physics, and molecular structures. Despite early risks, the development of radiation safety measures ensured their widespread adoption, solidifying their place as a cornerstone of modern technology. Röntgen's work not only earned him the first Nobel Prize in Physics but also inspired future advancements in imaging, from CT scans to MRI, continuing to illuminate the hidden layers of the world.

Next Chapter

While Röntgen's discovery of X-rays revealed the unseen within the human body, another breakthrough was brewing in the study of molecular structure. In the next chapter, we will explore how **Rosalind Franklin** used X-ray diffraction to capture the first images of DNA's helical structure. Her work, though overlooked at the time, would become central to one of the greatest discoveries in biology: the double-helix model of DNA, the blueprint of life.

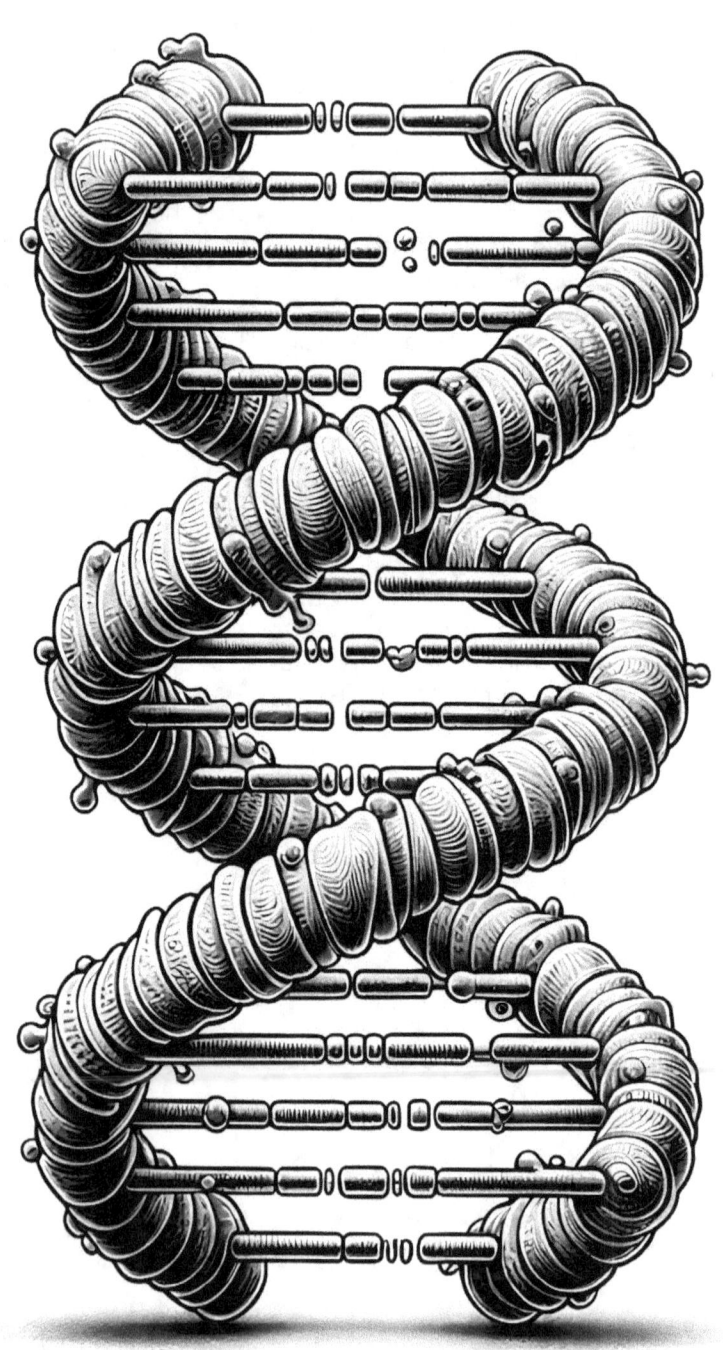

018 Seeing the Double Helix
An Overlooked Photo and DNA's Structure

In the early 20th century, the mystery of how genetic information was stored and transmitted was one of the most pressing questions in biology. While scientists had determined that **DNA (deoxyribonucleic acid)** was a critical component of cells, its precise structure remained unknown. Many believed that understanding DNA's form would unlock the secrets of heredity, mutation, and evolution. However, the tools and techniques needed to visualize molecules at such a microscopic scale were still in their infancy.

In the early 1950s, **Rosalind Franklin**, a brilliant and meticulous crystallographer at King's College London, was at the forefront of this effort. Franklin specialized in **X-ray diffraction**, a technique that allowed scientists to deduce the structure of molecules based on the patterns created when X-rays were scattered by crystalline substances. Her work on DNA, alongside her colleague Raymond Gosling, was groundbreaking. Franklin and Gosling captured an image known as **Photo 51**, which revealed a clear X-shaped diffraction pattern—an unmistakable clue that DNA had a **helical structure**.

While Franklin's data was extraordinarily precise, she was cautious about interpreting it prematurely. She meticulously analyzed her findings, aiming to construct a model based on solid evidence rather than speculation. Meanwhile, at the Cavendish Laboratory in Cambridge, **James Watson** and **Francis Crick** were racing to solve DNA's structure, relying on both their own theoretical work and data from other researchers. Watson and Crick obtained access to Photo 51 through Franklin's colleague **Maurice Wilkins**, without her knowledge or consent. This photograph provided the critical insight they needed to refine their theoretical models and propose the now-famous **double-helix structure of DNA**.

In April 1953, Watson and Crick published their groundbreaking paper in Nature, describing DNA as a pair of helical strands coiled around each other. While their model quickly gained acclaim, Franklin's crucial contributions were largely overlooked. Her detailed data and Photo 51 had been pivotal in confirming the helical structure, yet she received little recognition during her lifetime. Franklin's meticulous work, later acknowledged posthumously, remains a cornerstone in the story of DNA's discovery.

Importance and Impact

The determination of DNA's double-helix structure was one of the most significant scientific achievements of the 20th century, transforming biology and medicine. The discovery revealed how genetic information is stored, replicated, and transmitted. Watson and Crick's model demonstrated that DNA's two strands are held together by **base pairs**—adenine pairing with thymine, and cytosine with guanine—forming a template for replication. This insight provided the first glimpse into the **molecular basis of heredity**, explaining how genetic traits are passed from one generation to the next.

Franklin's X-ray diffraction work was instrumental in this breakthrough. Without the clarity of Photo 51, Watson and Crick may not have confirmed their model so quickly. Franklin's contributions extended beyond the double helix; she also conducted critical work on the **structural properties of RNA and viruses**, paving the way for advancements in molecular biology and virology. Her data, while underappreciated during her life, became foundational in understanding DNA as the molecule of life.

The double-helix discovery also ushered in the era of **molecular biology**, enabling scientists to study genes and their functions at the molecular level. This led to significant breakthroughs, including the identification of genetic mutations responsible for diseases, the development of genetic engineering techniques, and the sequencing of the human genome. These advancements transformed fields ranging from medicine to agriculture, enabling targeted therapies, genetic diagnostics, and the development of genetically modified organisms (GMOs).

Franklin's story became emblematic of the challenges faced by women in science, particularly during the mid-20th century. Her exclusion from the Nobel Prize in 1962, awarded to Watson, Crick, and Wilkins for their work on DNA, underscored the systemic marginalization of female scientists. Franklin's legacy has since been reexamined, and she is now celebrated as a pioneer whose meticulous research contributed to one of biology's greatest discoveries.

The discovery of DNA's structure was not only a scientific triumph but also a cultural milestone, fundamentally altering humanity's understanding of life. It inspired philosophical and ethical debates about genetics and human identity, from questions about genetic determinism to concerns about the implications of genetic engineering. The double helix, immortalized in textbooks and scientific iconography, became a symbol of modern biology and the quest to understand life at its most fundamental level.

Chapter Summary

The discovery of DNA's double-helix structure revolutionized biology, providing the first clear understanding of how genetic information is stored and replicated. Rosalind Franklin's Photo 51 was instrumental in confirming the helical structure, though her contributions went largely unrecognized during her lifetime. Watson and Crick's model revealed the molecular basis of heredity, laying the foundation for molecular biology and enabling transformative advancements in genetics, medicine, and biotechnology. Franklin's meticulous work remains a cornerstone of this historic achievement, and her legacy serves as a reminder of the collaborative nature of science and the importance of recognizing all contributors.

Next Chapter

With the structure of DNA revealed, scientists began exploring the practical applications of this knowledge. One of the most significant breakthroughs in applying biological insights came with the invention of the **pacemaker**, a device that uses electrical impulses to regulate the heart's rhythm. In the next chapter, we'll explore how a simple misconnection in an electrical circuit led to the creation of a life-saving technology, bridging biology and engineering in an extraordinary way.

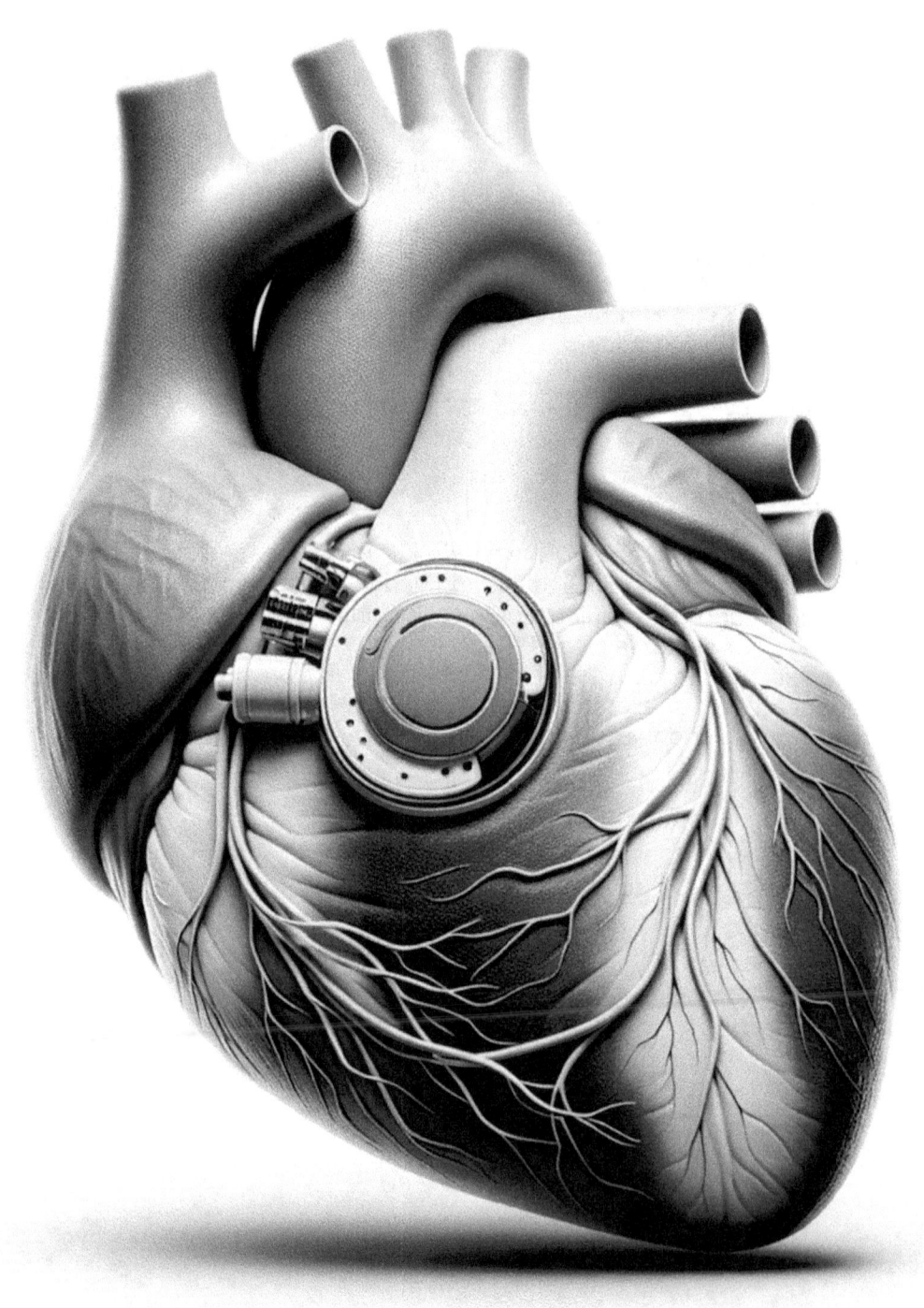

019 The Pacemaker's Pulse
A Misconnected Circuit for Heartbeats

In the mid-20th century, advances in **medicine and electronics** were converging to address critical challenges in human health. Among these challenges was the problem of **cardiac arrhythmias**—irregular heartbeats that could cause severe complications or even death. While physicians had identified the issue, they lacked reliable methods to manage or correct it in real time. At the time, doctors had begun experimenting with rudimentary devices to regulate heartbeats, but these were large, cumbersome, and required direct access to electrical outlets. The idea of a portable device to regulate the heart's rhythm seemed out of reach.

In 1958, Canadian engineer **John Hopps** and American cardiologist **Dr. William Greatbatch** were independently working on technologies to address this problem. It was during this period that Greatbatch made an accidental discovery that would revolutionize cardiac care. While building a circuit for a device to record heart sounds, Greatbatch mistakenly inserted the wrong **electronic resistor** into the circuit. When testing the circuit, he noticed it emitted regular, rhythmic electrical pulses—mimicking the natural pacing of the human heart. This error sparked the realization that such a device could potentially stimulate a failing heart to beat regularly.

Encouraged by his serendipitous discovery, Greatbatch began refining his invention. He worked tirelessly to develop a compact, reliable device that could generate precise electrical pulses to regulate heartbeats. After rigorous testing, Greatbatch collaborated with a surgical team to implant the first fully functional **implantable pacemaker** into a human patient. The success of this groundbreaking procedure proved that technology could directly intervene in the body's most essential biological processes, opening a new frontier in **biomedical engineering**.

Importance and Impact

The invention of the **pacemaker** marked a major leap forward in medicine and technology. For patients with arrhythmias or heart block, the pacemaker transformed what was once a fatal condition into a manageable one, vastly improving quality of life and survival rates. Greatbatch's device demonstrated that **electronics and biology** could work together in unprecedented ways, setting the stage for a new era of **cybernetic medicine**—the integration of machines with human physiology.

The first pacemakers were relatively basic by today's standards. They delivered fixed electrical pulses to the heart, with no ability to adjust to the patient's activity level. However, these early devices were revolutionary in their time, proving the concept and inspiring rapid innovation. Over the next few decades, engineers developed **programmable pacemakers** capable of sensing the body's needs and adjusting their pacing accordingly. Today's pacemakers are compact, durable, and powered by advanced lithium batteries, allowing them to function for years without replacement.

The broader implications of the pacemaker extended beyond cardiology. The success of this device catalyzed the development of other implantable technologies, including **defibrillators, cochlear implants**, and **neurostimulators**. Each of these innovations built upon the idea that electrical signals could be harnessed to repair or enhance biological functions. Greatbatch's work became a template for the growing field of **bioelectronics**, where engineering and medicine collaborate to address previously untreatable conditions.

In addition to its medical impact, the pacemaker highlighted the power of **serendipity in innovation**. Greatbatch's accidental resistor mishap underscored how unexpected errors could lead to groundbreaking discoveries. His story inspired engineers and scientists to embrace creativity, curiosity, and persistence in their work, even when facing failure or uncertainty. The pacemaker also encouraged collaboration across disciplines, uniting doctors, engineers, and researchers in the pursuit of life-saving technologies.

The pacemaker's cultural impact was equally profound. It shifted public perceptions of technology, demonstrating that machines could not only augment but also sustain life. Patients who once lived under the constant threat of heart failure could now look forward to longer, healthier lives. By creating a direct link between electronics and human health, the pacemaker paved the way for a future where the boundaries between biology and technology continue to blur.

Chapter Summary

Dr. William Greatbatch's accidental discovery of the pacemaker revolutionized medicine, providing a life-saving solution for patients with irregular heart rhythms. By harnessing the power of electrical pulses to regulate the heart, the pacemaker not only transformed cardiac care but also set the stage for the field of bioelectronics. This innovation demonstrated the potential of integrating machines with biological systems, inspiring a wave of medical technologies that continue to shape modern healthcare. The pacemaker remains a testament to the power of curiosity, interdisciplinary collaboration, and the serendipity that drives scientific progress.

Next Chapter

As the pacemaker illustrated the ability of technology to work in harmony with biology, another innovation drew inspiration directly from nature. In the next chapter, we'll explore how the structure of **silkworm silk** and **spider webs** inspired the development of **biosteel**, a lightweight, high-strength material with applications ranging from medicine to aerospace. This leap in biomimicry would showcase how nature's designs continue to guide humanity's most advanced innovations.

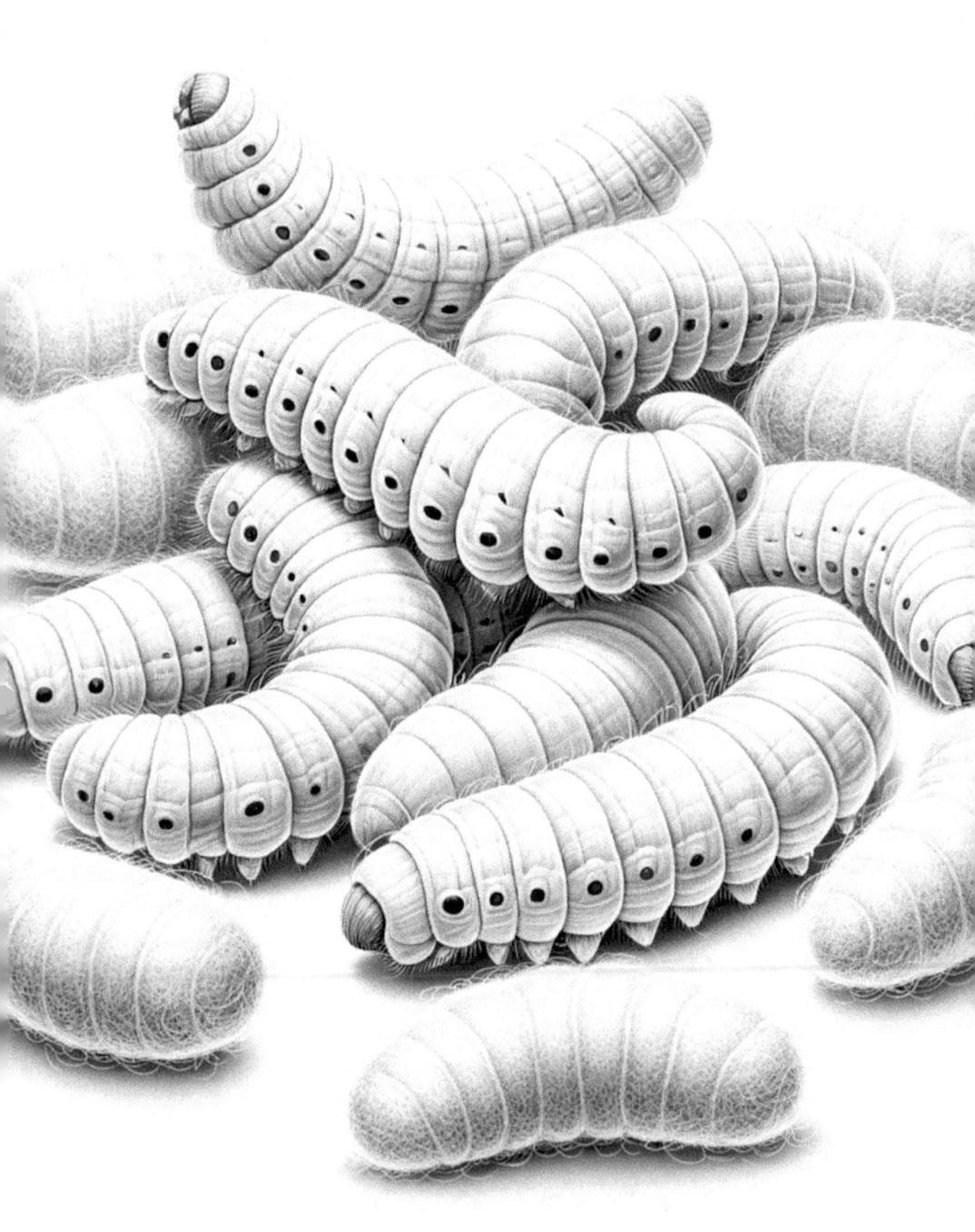

Silkworm's Steel

Nature's Inspiration for Biosteel

Throughout history, **nature** has served as a source of inspiration for scientific and technological advancements. One of the most intriguing natural materials is **spider silk**, renowned for its extraordinary properties. Spider silk is both **stronger than steel** by weight and highly flexible, making it an ideal candidate for use in advanced materials. However, spiders cannot be farmed efficiently due to their territorial and cannibalistic behavior, limiting the availability of this remarkable substance. In the 20th century, researchers began to wonder: Could the properties of spider silk be replicated in the lab or by other means?

This question led to experiments in **biomimicry**, a field that seeks to replicate or adapt natural phenomena for human use. Scientists turned their attention to **silkworms**, creatures that produce silk for their cocoons, which is weaker than spider silk but produced in greater quantities. By the late 1990s, a team led by **Dr. Randy Lewis**, a biologist, had an extraordinary idea: What if silkworms could be genetically engineered to produce silk with the properties of spider silk? This breakthrough would require a deep understanding of genetics and protein engineering.

Lewis and his team extracted the **genes responsible for spider silk production** and inserted them into the DNA of silkworms. The result was a genetically modified silkworm capable of producing a hybrid silk—one that combined the strength and elasticity of spider silk with the mass-production capabilities of silkworms. This hybrid material, later dubbed **"biosteel"**, was hailed as a breakthrough in material science. The silk could potentially be used in applications ranging from **medical sutures** and **bulletproof vests** to **aerospace materials** and **artificial tendons**.

Importance and Impact

The invention of biosteel represented a **milestone in biomaterials research**, offering a glimpse into how biological systems could be harnessed to create new materials. Biosteel demonstrated the power of **genetic engineering** and biomimicry, showing that by understanding and modifying natural processes, humans could create materials with unprecedented properties. The potential applications of biosteel were wide-ranging. In **medicine**, it offered a way to create stronger, more biocompatible sutures, ligaments, and artificial tissues. In **military** and

aerospace industries, biosteel promised lightweight yet incredibly durable materials for armor, parachutes, and spacecraft.

Biosteel also inspired advancements in **sustainable manufacturing**. Traditional materials like steel or synthetic fibers require energy-intensive processes and often rely on non-renewable resources. By contrast, biosteel production relies on silkworms, a renewable and relatively low-energy source. This bio-based approach demonstrated the feasibility of producing high-performance materials in a way that could reduce the environmental impact of manufacturing.

Despite its promise, biosteel faced challenges in commercialization. Producing spider-silk proteins in large enough quantities proved to be technically difficult, and the cost of scaling up production remained high. However, this limitation did not diminish its impact on scientific research. Biosteel inspired a wave of innovation in **synthetic biology**, leading to the development of other bioengineered materials, such as artificial leather, plant-based plastics, and bioadhesives. It also paved the way for researchers to explore **CRISPR technology** and more precise genetic editing tools to improve material properties further.

Biosteel's success underscored the importance of looking to nature for solutions to complex problems. Biomimicry has since become a cornerstone of material science, influencing the design of everything from **self-cleaning surfaces** (inspired by lotus leaves) to **lightweight building materials** (modeled after bone structures). By mimicking the strength, flexibility, and efficiency of spider silk, biosteel demonstrated how evolutionary solutions could address modern technological challenges, providing a sustainable alternative to traditional manufacturing methods.

Chapter Summary

The creation of biosteel, inspired by spider silk and produced through genetically engineered silkworms, marked a breakthrough in material science and genetic engineering. By combining the natural strength and flexibility of spider silk with the mass-production capabilities of silkworms, researchers developed a material with wide-ranging applications in medicine, aerospace, and sustainability. Though commercial challenges remain, biosteel has influenced the fields of biomimicry and synthetic biology, inspiring a new era of eco-friendly and high-performance materials. This achievement underscores how nature's designs can guide scientific innovation and address humanity's technological needs.

Next Chapter

As researchers explored the potential of bioengineered materials, the study of the **fundamental components of matter** took a dramatic leap forward. The next chapter delves into the discovery of the **electron**, a breakthrough that not only transformed our understanding of atomic structure but also laid the groundwork for the field of quantum mechanics and the development of modern electronics. Join us as we step into the lab of **J.J. Thomson**, whose experiments with cathode rays revealed one of nature's most fundamental particles.

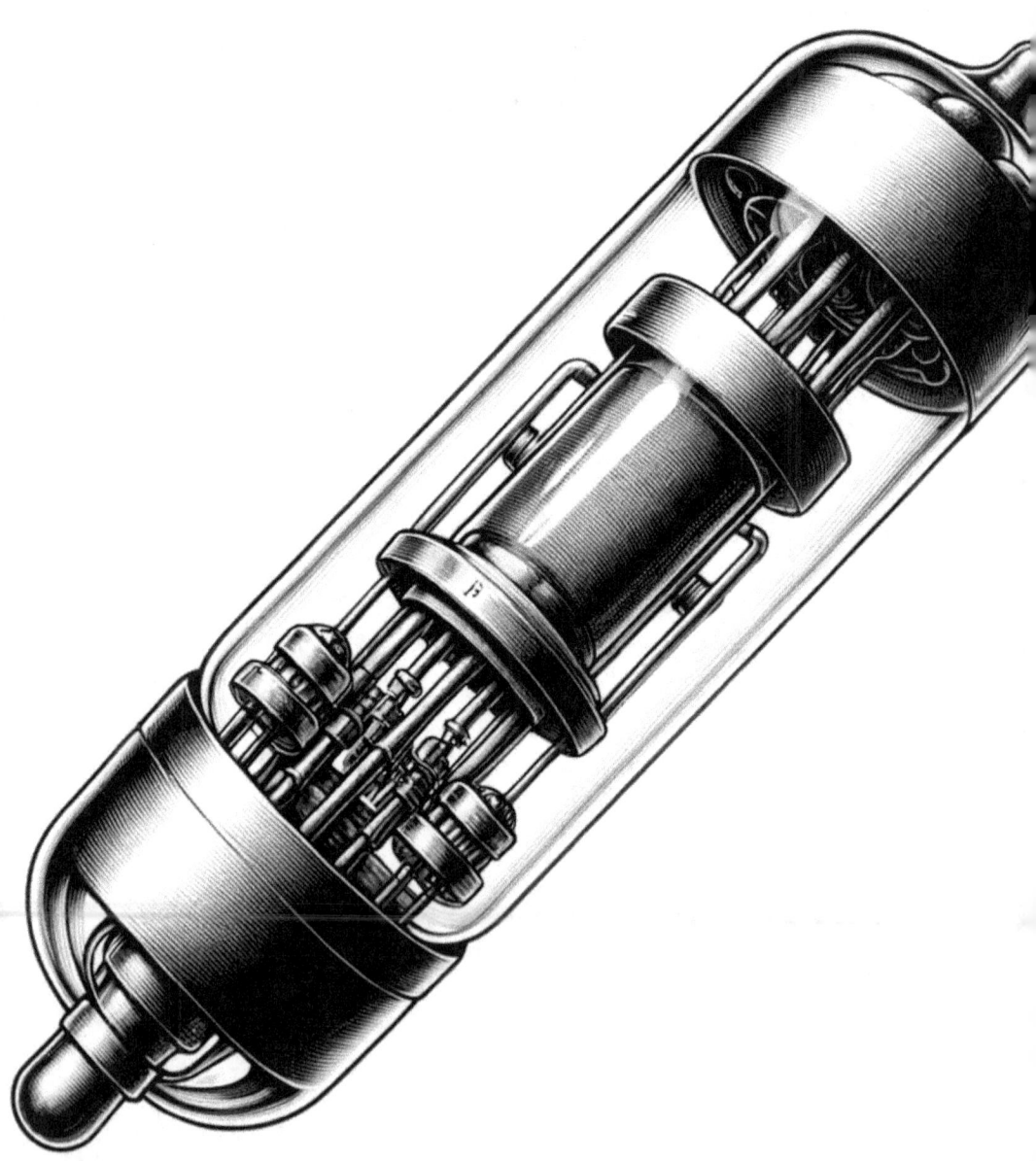

021 Discovering the Electron
Shadows of Light and Cathode Rays

By the late 19th century, the understanding of matter was undergoing a profound transformation. While the concept of atoms had existed since antiquity, their structure remained a mystery. Most scientists still viewed atoms as indivisible units of matter. However, as technology advanced, experiments with **electricity and gases** began to hint that atoms might contain smaller, fundamental components. Among the scientists intrigued by these possibilities was **J.J. Thomson**, a physicist working at the **Cavendish Laboratory** at the University of Cambridge.

Thomson's interest was drawn to the mysterious phenomena of **cathode rays**, first observed in the mid-19th century. These rays were produced when an electric current passed through a **vacuum tube** with electrodes at each end. The rays emitted from the negative electrode (the cathode) seemed to travel in straight lines, creating a faint green glow when they struck the tube's inner surface. Early experiments suggested that cathode rays carried energy, but their nature was fiercely debated. Some believed they were a form of **light**, while others theorized they were **particles**.

In 1897, Thomson designed a series of experiments to settle the question. Using a specially modified cathode ray tube, he placed **electromagnetic fields** around the tube and observed how the rays behaved. To his surprise, the cathode rays bent under the influence of the fields, indicating they were **charged particles**, not waves of light. Moreover, the particles always carried a **negative charge** and had the same properties regardless of the type of gas in the tube. This consistency suggested that the particles were a fundamental component of all matter. Thomson had discovered what would later be named the **electron**, the first subatomic particle.

Importance and Impact

The discovery of the electron shattered the prevailing view of the atom as an indivisible unit and marked the birth of **modern atomic physics**. Thomson's experiments revealed that atoms were not solid, indivisible entities but were instead composed of smaller components. This breakthrough fundamentally changed how scientists viewed matter and led to the development of new models of the atom. Thomson proposed the **"plum pudding model"**, which envisioned atoms as spheres of positive charge with negatively charged electrons embedded within them, like raisins in a pudding. Though this model would later be replaced, it represented a critical step toward understanding atomic structure.

The discovery of the electron also had profound implications for the emerging field of **electromagnetism**. Scientists quickly realized that electrons played a crucial role in **electricity** and **magnetism**, forming the basis of electrical conductivity and chemical bonding. This understanding enabled advancements in technologies such as **electric motors**, **generators**, and **telecommunication systems**, which relied on manipulating the flow of electrons. Thomson's work bridged the gap between experimental physics and practical applications, accelerating the development of electrical and electronic technologies that would define the 20th century.

Beyond its immediate applications, Thomson's discovery laid the groundwork for a series of transformative scientific advancements. It inspired **Ernest Rutherford's experiments** on the structure of the atom, which revealed the nucleus, and eventually led to the development of **quantum mechanics**, a field that explains the behavior of particles at the atomic and subatomic levels. These advancements would revolutionize physics, chemistry, and materials science, driving innovations in everything from semiconductors to nuclear energy.

Thomson's discovery of the electron also opened the door to the **chemical industry's transformation**. By revealing the atom's internal structure, it became possible to understand and manipulate the behavior of molecules, spurring the rise of synthetic chemistry. This paved the way for the creation of entirely new materials, such as **plastics**, which would soon revolutionize manufacturing, engineering, and everyday life. The electron's discovery thus connected physics to chemistry in a way that reshaped industries and technologies worldwide.

Thomson's contributions did not go unrecognized in his lifetime. In 1906, he was awarded the **Nobel Prize in Physics** for his groundbreaking work on the electron. Many of his students, including **Ernest Rutherford** and **Niels Bohr**, would go on to make their own monumental discoveries, cementing Thomson's legacy as a central figure in the development of modern science. The electron's discovery was not merely a scientific triumph but a transformative event that touched nearly every aspect of modern life, from theoretical research to industrial applications.

Chapter Summary

J.J. Thomson's discovery of the electron in 1897 revolutionized the understanding of matter, marking the first step in revealing the atom's internal structure. His experiments with cathode rays demonstrated that atoms were not indivisible but contained smaller, negatively charged particles. This breakthrough laid the foundation for modern atomic theory, quantum mechanics, and the development of electrical and electronic technologies. Thomson's work bridged physics and chemistry, driving advancements that reshaped industries, from telecommunications to synthetic materials. The electron's discovery transformed science, enabling a new understanding of matter and sparking a wave of innovation that continues to this day.

Next Chapter

With the discovery of the electron opening the door to a deeper understanding of atomic and molecular behavior, scientists began to explore ways to manipulate matter at the molecular level. In the next chapter, we'll follow the story of **Bakelite**, the world's first synthetic plastic, and how it revolutionized materials science. This accidental discovery would mark the dawn of the **plastic age**, transforming industries and everyday life by introducing a material that could be molded into virtually any shape.

022 Plastic's First Step
Bakelite and the Birth of Synthetics

At the turn of the 20th century, industries were on the lookout for **new materials** to meet the demands of the rapidly advancing technological age. Traditional materials like wood, metals, and natural fibers had their limitations—they were often expensive, heavy, or scarce. The search for lightweight, durable, and versatile materials led scientists and engineers to explore the potential of **synthetic compounds**, especially those derived from chemical reactions. One of the most important breakthroughs in this field came in 1907 when **Leo Baekeland**, a Belgian-American chemist, invented **Bakelite**, the first fully synthetic plastic.

Baekeland, originally known for inventing Velox photographic paper, turned his attention to the burgeoning electrical and automotive industries, which needed materials that were **heat-resistant** and could act as insulators. Inspired by earlier experiments with phenol and formaldehyde, Baekeland sought to create a synthetic resin that could harden under heat and pressure. After a series of meticulous experiments, Baekeland succeeded in synthesizing a hard, moldable, and non-conductive material, which he patented as **Bakelite**.

What made Bakelite revolutionary was its ability to **retain its shape and properties** after being heated and set, a process now known as **thermosetting**. Unlike natural materials, which often degrade or warp under extreme conditions, Bakelite was stable, durable, and resistant to heat and chemicals. These properties made it an ideal material for electrical insulators, radio casings, and automobile parts, all of which were critical to the industries driving the 20th century's technological revolution.

Importance and Impact

The invention of Bakelite marked the beginning of the **Age of Plastics**, fundamentally changing how materials were used in manufacturing and everyday life. For the first time, industries had access to a material that could be **mass-produced, customized**, and **molded into virtually any shape**. This flexibility allowed Bakelite to replace traditional materials in a wide range of applications, from electrical components to household items like combs, buttons, and even jewelry. Bakelite's versatility made it an essential material for industries such as **telecommunications, automotive manufacturing**, and **consumer goods**.

The impact of Bakelite went beyond its immediate applications. It represented the first successful step into the world of **synthetic polymers**, paving the way for the development of other plastics such as polystyrene, polyvinyl chloride (PVC), and nylon. These newer plastics built upon Baekeland's innovations, offering even greater versatility, strength, and durability. By enabling the creation of lightweight, inexpensive, and mass-producible materials, Bakelite contributed to the rise of **modern consumer culture**, making once-luxurious goods accessible to a broader population.

Bakelite's properties also opened up new possibilities in **design** and **aesthetics**. Its ability to mimic natural materials like wood, ivory, and tortoiseshell allowed it to become a popular material for decorative and artistic purposes. During the **Art Deco era**, Bakelite was used to craft elegant, colorful objects, from radios to jewelry, showcasing its aesthetic appeal in addition to its functional qualities. This dual role—both utilitarian and decorative—cemented Bakelite's place in industrial design and everyday life.

Bakelite's invention also sparked discussions about the **environmental implications** of synthetic materials. While it initially solved issues of material scarcity, its durability—once celebrated—eventually raised concerns about waste and pollution. As newer plastics were developed and proliferated, the challenge of managing non-biodegradable materials became increasingly apparent. Today, while Bakelite itself is no longer widely used, its legacy remains a reminder of how innovations can have unintended long-term consequences.

Moreover, Bakelite's creation underscored the importance of **applied chemistry** in driving technological innovation. Baekeland's success demonstrated how chemistry could be harnessed to create entirely new materials with properties tailored to industrial needs. This achievement inspired further research into synthetic materials, shaping industries ranging from construction to medicine and influencing fields like polymer science, chemical engineering, and industrial design.

Chapter Summary

Leo Baekeland's invention of Bakelite in 1907 ushered in the era of **synthetic plastics**, transforming manufacturing, design, and everyday life. As the first fully synthetic material, Bakelite offered durability, heat resistance, and moldability, meeting the demands of rapidly growing industries such as telecommunications and automotive manufacturing. Its versatility and mass-producibility made it a cornerstone of modern industrialization and consumer culture. Bakelite also laid the foundation for the development of other plastics, which would become essential to nearly every facet of modern life. Though it solved many material challenges of its time, Bakelite also foreshadowed the environmental dilemmas posed by synthetic materials, highlighting the complex legacy of innovation.

Next Chapter

As Bakelite revolutionized material science, other accidental discoveries were also reshaping industries in unexpected ways. In the next chapter, we'll explore how a melted candy bar in a radar technician's pocket led to the discovery of **microwave heating** and the invention of the **microwave oven**. This serendipitous breakthrough would transform cooking, medicine, and communications, showing once again how chance plays a critical role in technological progress.

023 The Microwave Oven Mistake
A Candy Bar and the Advent of Microwave Cooking

The 1940s marked a time of intense technological innovation, much of it driven by the demands of **World War II**. Among these innovations was the development of **radar technology**, used to detect enemy aircraft and ships. Engineers and scientists worked tirelessly to refine radar systems, often encountering unexpected phenomena along the way. One such moment of serendipity occurred in 1945, when **Percy Spencer**, an American engineer working for the Raytheon Corporation, stumbled upon a discovery that would change the way people cooked forever.

Spencer, a self-taught engineer and expert in radar technology, was working on a **magnetron**, a critical component of radar systems that generates high-frequency microwaves. While testing a magnetron, Spencer noticed something peculiar: **a candy bar in his pocket had melted**. Intrigued, he began experimenting to understand why. Spencer placed popcorn kernels near the magnetron, and to his amazement, they began to pop. His experiments continued, this time with an egg, which promptly exploded under the intense heat generated by the microwaves.

What Spencer had discovered was that **microwaves could heat food** by causing water molecules within the food to vibrate rapidly, creating friction and producing heat. This phenomenon was entirely unintentional—an accidental byproduct of radar research—but it opened up a new realm of possibilities for cooking technology. Spencer realized the potential of this discovery and began developing a device specifically designed to harness microwaves for cooking.

By 1947, Raytheon had introduced the first commercial **microwave oven, the Radarange**. However, these early models were far from practical for home use—they were massive, expensive, and primarily used in industrial settings like restaurants and military kitchens. It wasn't until the 1960s and 1970s, with advancements in miniaturization and cost reduction, that microwave ovens became accessible to the average household, revolutionizing how people cooked and reheated food.

Importance and Impact

Percy Spencer's accidental discovery of microwave cooking transformed the food industry and revolutionized domestic life. The invention of the **microwave oven** provided a new, efficient way to prepare food, dramatically reducing cooking times and enabling busy households to keep up with modern life's fast pace. Foods that once took hours to prepare could now be cooked in minutes, making the microwave oven an indispensable appliance in kitchens worldwide.

Beyond its impact on cooking, Spencer's discovery had **scientific and industrial implications**. Microwave technology became a cornerstone of **medical devices**, such as diathermy machines used for therapeutic heating in physical therapy. It also found applications in materials processing, helping industries cure adhesives, dry ceramics, and even produce foam rubber. The principles behind microwave heating also influenced communications technology, as microwaves were already being harnessed in radar and later became integral to satellite and cellular communication systems.

The invention of the microwave oven also reshaped **consumer habits and the food industry**. Pre-packaged microwaveable meals became a staple in grocery stores, altering how food was produced, marketed, and consumed. The popularity of the microwave oven spurred the development of **microwave-safe containers and packaging materials**, further expanding its applications. The microwave also contributed to changing cultural norms, particularly in how people approached convenience and time management, reinforcing the notion of fast, on-demand solutions in daily life.

However, the introduction of the microwave oven wasn't without challenges. Early adopters were skeptical about its safety, with concerns that microwave radiation might pose health risks. Education campaigns and scientific studies eventually alleviated these fears, but public apprehension highlighted the challenges of introducing disruptive technologies. Despite these hurdles, the microwave oven gained widespread acceptance and became a symbol of modern convenience, a testament to the transformative power of innovation born from unexpected moments.

Spencer's discovery also underscores the importance of **curiosity and experimentation** in scientific progress. His willingness to explore an odd phenomenon—a melted candy bar—led to an invention that transformed domestic life and the food industry. It serves as a reminder that some of the most impactful discoveries arise not from planned research but from a keen eye for the unexpected and a determination to pursue it.

Chapter Summary

Percy Spencer's accidental discovery of microwave cooking in 1945, while working on radar technology, marked the birth of the **microwave oven**. What began with a melted candy bar evolved into a transformative appliance that revolutionized cooking and reshaped the food industry. Spencer's ingenuity turned an unexpected observation into a practical device that saved time, changed eating habits, and influenced industries ranging from medicine to communications. The microwave oven's story is a testament to the role of serendipity in innovation, highlighting how unexpected moments can lead to groundbreaking advancements that redefine everyday life.

Next Chapter

As microwave technology found its way into kitchens and industries, other scientific discoveries continued to shape how humans interacted with the natural world. In the next chapter, we'll explore the invention of **radar for meteorology**, where wartime radar systems were adapted to detect weather patterns. This unexpected application of radar would revolutionize weather forecasting, helping to predict storms and save countless lives in the process.

024 Radar for Rain
How Wartime Tech Predicted Weather

During **World War II**, radar technology became one of the most critical tools for military operations, enabling the detection of enemy aircraft and ships long before they were visible to the naked eye. Radar, which stands for **Radio Detection and Ranging**, worked by sending out radio waves and analyzing the signals that bounced back from objects. By the 1940s, scientists and engineers had refined radar systems to an extraordinary degree, making them invaluable on the battlefield. However, as with many wartime technologies, radar's potential would soon find applications far beyond its original purpose.

One of the most surprising discoveries during radar's development occurred when operators noticed unexpected "blips" on their screens that did not correspond to aircraft or ships. These anomalies were initially seen as nuisances, but scientists quickly realized that they were caused by **precipitation**, such as rain, snow, or even hail. At first, this observation was dismissed as a distraction from radar's primary purpose, but a few curious researchers began to investigate further. They hypothesized that radar could be used to detect and measure **weather patterns**, offering a new way to study and predict the movement of storms.

After the war, military radar systems were adapted for meteorological purposes, marking the birth of **weather radar**. In 1947, researchers began using surplus radar equipment to track storms and precipitation patterns systematically. One of the first significant applications came in the early 1950s, when weather radar was used to monitor **hurricane movements**, providing life-saving data for coastal communities. By the late 20th century, radar had become an indispensable tool for meteorologists, enabling accurate weather predictions and advanced warnings of severe weather events.

Importance and Impact

The adaptation of radar for meteorology revolutionized the way humans understood and prepared for **weather phenomena**. Before radar, weather forecasting relied heavily on ground observations, barometric pressure readings, and rudimentary models. While these methods provided some insights, they were often imprecise and lacked the ability to track fast-moving storms. Weather radar changed this by providing a **real-time view of precipitation**, allowing meteorologists to monitor storm intensity, movement, and structure with unprecedented accuracy.

This advancement had profound implications for public safety. Weather radar made it possible to issue early warnings for **severe storms**, **tornadoes**, and **hurricanes**, saving countless lives by giving people more time to seek shelter. For example, radar's ability to detect tornadoes in their early stages has significantly reduced fatalities in tornado-prone regions. Similarly, radar-based hurricane tracking has allowed for more effective evacuations and disaster planning, minimizing the loss of life and property in vulnerable coastal areas.

Radar's impact extended beyond safety, influencing industries such as **agriculture**, **aviation**, and **transportation**. Farmers began using weather radar to plan planting and harvesting schedules, reducing losses caused by unexpected storms. Airlines relied on radar to navigate safely through turbulent weather, and shipping companies used it to avoid storms at sea. These applications not only improved efficiency but also reduced economic losses across multiple sectors.

In addition to its practical applications, radar opened up new frontiers in **atmospheric science**. By providing detailed images of precipitation patterns, radar enabled scientists to study the dynamics of storms, clouds, and atmospheric processes in ways that were previously impossible. This led to a deeper understanding of phenomena like **thunderstorms**, **cyclones**, and **monsoons**, paving the way for advances in climate science and weather modeling. Today's sophisticated radar systems, including Doppler radar, are direct descendants of these early innovations, capable of detecting wind patterns and even the rotation of tornadoes.

The story of radar's adaptation for meteorology also underscores the importance of **serendipity in science**. What began as an unintended side effect of wartime technology became a transformative tool for understanding the natural world. This transition highlights how curiosity and a willingness to explore unexpected results can lead to breakthroughs that benefit society in profound ways.

Chapter Summary

The discovery that radar could detect precipitation transformed the technology from a wartime tool into a cornerstone of modern meteorology. By enabling real-time tracking of storms and other weather patterns, radar revolutionized weather forecasting, improving public safety and enhancing industries such as aviation, agriculture, and transportation. It also advanced atmospheric science, providing new insights into the dynamics of storms and climate systems. The adaptation of radar for meteorology is a testament to the power of curiosity and the unforeseen benefits of scientific innovation.

Next Chapter

As radar technology transformed the study of the skies, another accidental discovery was quietly reshaping life on the ground. In the next chapter, we'll explore the invention of **Velcro**, a hook-and-loop fastener inspired by nature. This simple yet ingenious invention, born from a chance observation during a hike, would revolutionize industries from fashion to space exploration, showcasing the potential of biomimicry in design.

025 The Velcro Hike

Burrs and the Invention of Hook-and-Loop Fasteners

The natural world has always inspired inventors, but few stories of biomimicry are as impactful—and as unexpected—as the invention of **Velcro**. In 1941, **George de Mestral**, a Swiss engineer, embarked on a hike through the Alpine countryside with his dog. When he returned home, he noticed that both his clothing and his dog's fur were covered with **burrs**, the small seed pods of burdock plants. Rather than simply brushing them off in frustration, de Mestral's curiosity got the better of him. He decided to examine the burrs under a microscope, hoping to understand how they clung so tightly to fabric and fur.

What he observed was a **natural marvel of design**. Each burr was covered in tiny hooks that easily latched onto the loops found in fibers and animal hair. This simple but effective mechanism inspired de Mestral to replicate the process artificially. He envisioned a fastener that mimicked the hook-and-loop structure he had seen in the burrs, one that could bind materials together quickly, securely, and without the need for buttons, zippers, or ties.

Over the next several years, de Mestral worked tirelessly to perfect his invention. Using **nylon**, a relatively new synthetic material, he created a system of tiny hooks and loops that could be pressed together to form a bond and then easily separated. In 1955, he patented his creation, naming it **Velcro**, a portmanteau of the French words **"velours"** (**velvet**) and **"crochet"** (**hook**). At first, Velcro was met with skepticism and slow adoption, but it eventually found its footing in niche applications and, later, global markets.

Importance and Impact

Velcro's design revolutionized the concept of fasteners, offering a **versatile**, **reusable**, and **easy-to-use** alternative to traditional methods. Its durability and simplicity made it ideal for a wide range of applications, from securing shoes and clothing to medical devices and industrial equipment. Velcro's utility shone brightest in situations where speed, convenience, or precision were crucial, making it an indispensable tool in numerous industries.

One of Velcro's most famous applications came in the **space program**. In the 1960s, NASA adopted Velcro for use in the Apollo missions, where it became essential for keeping tools, equipment, and even food pouches secured in the weightless environment of space. The association with the space race helped boost Velcro's reputation and popularity, cementing its status as a cutting-edge innovation. Velcro also found its way into **military gear**, **medical braces**, and **adaptive clothing**, enabling improvements in functionality and accessibility.

In the world of fashion and consumer goods, Velcro became a household name, appearing on everything from children's shoes to luggage. It offered a solution for individuals who had difficulty using traditional fasteners, such as children or people with disabilities, thereby increasing accessibility and usability. Velcro's adaptability also made it a favorite among designers and engineers, who used it in creative ways to solve everyday problems.

The invention of Velcro is a prime example of **biomimicry**, where nature's designs inspire human innovation. George de Mestral's keen observation of a seemingly mundane phenomenon—a burr clinging to his clothing—underscored the power of curiosity and persistence. By turning a small inconvenience into a global innovation, de Mestral demonstrated how closely studying nature can lead to groundbreaking ideas. His work also inspired future inventors and scientists to look at nature not as an obstacle to be overcome but as a **blueprint for innovation**.

Today, Velcro continues to be a ubiquitous product with applications in nearly every industry. Its influence extends beyond its physical form; it symbolizes how simple observations, paired with determination, can lead to creations that transform everyday life. Velcro's legacy serves as a reminder that groundbreaking ideas often lie in the details of the ordinary, waiting for someone to notice.

Chapter Summary

George de Mestral's discovery of Velcro in the 1940s, inspired by the burrs that clung to his clothing, showcased the power of curiosity and nature-inspired design. Velcro's hook-and-loop fastening system revolutionized industries, finding uses in space exploration, fashion, medicine, and more. Its simplicity and versatility made it a global phenomenon, and its story highlighted the value of biomimicry in problem-solving. Velcro remains a testament to how careful observation and persistence can turn everyday nuisances into transformative innovations.

Next Chapter

While Velcro revolutionized fasteners through biomimicry, another scientific breakthrough would soon change how humanity viewed cellular structures. In the next chapter, we'll delve into the story of **electron microscopy** and how it allowed scientists to visualize cells and their intricate machinery for the first time. This leap in imaging technology would open up new frontiers in biology, medicine, and material science.

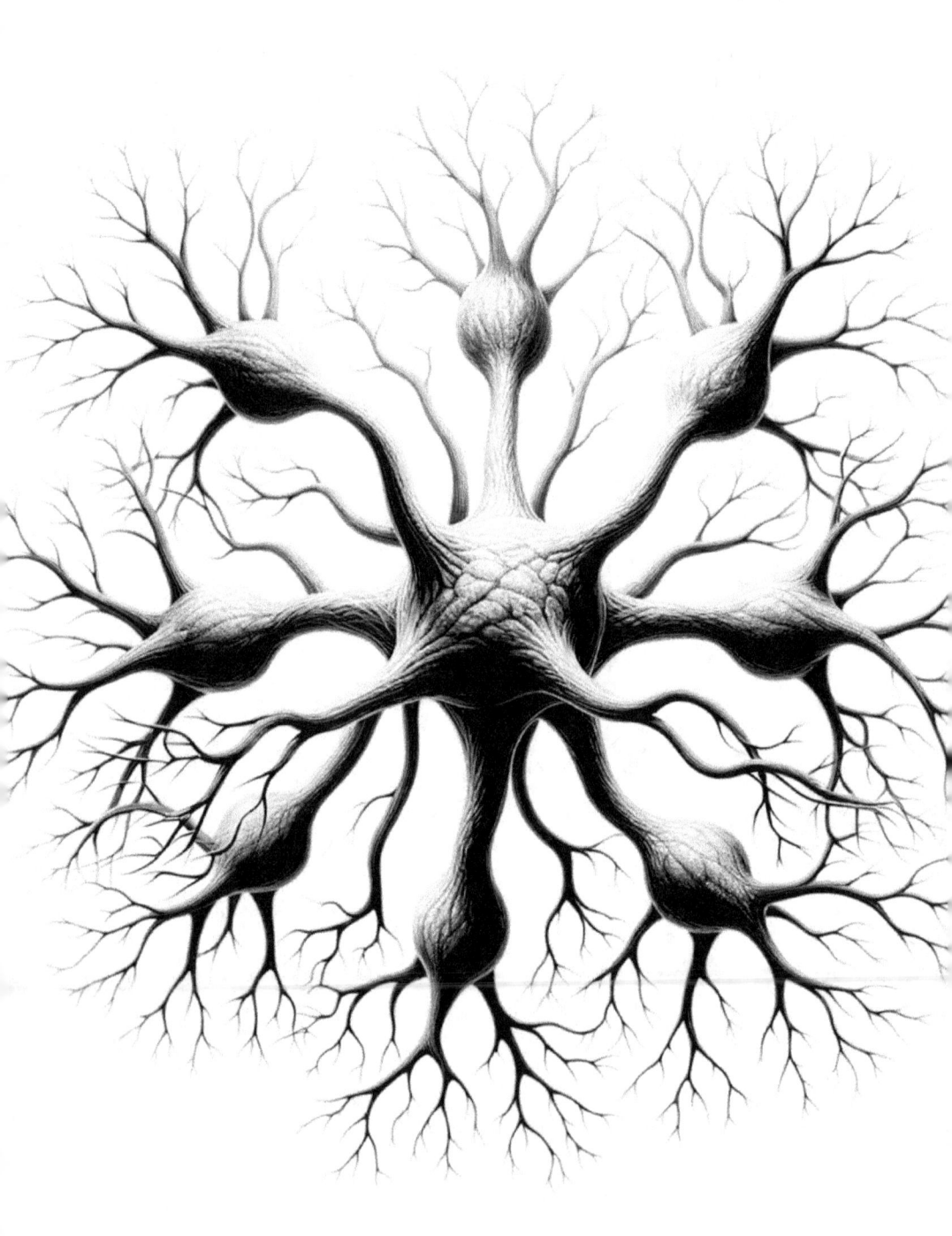

026 Neurons on Film
Electron Microscopy Reveals the Cell's Machinery

By the early 20th century, the **microscope** had already revolutionized biology, allowing scientists to explore the microscopic world in unprecedented detail. However, the limitations of optical microscopes were becoming increasingly apparent. Restricted by the wavelength of visible light, these microscopes could not resolve structures smaller than about **200 nanometers**, leaving much of the cellular world—especially organelles and macromolecular structures—beyond reach. Scientists knew that there was more to uncover, particularly within the intricate **machinery of cells**.

In the 1930s, the advent of **electron microscopy** offered a solution to this problem. Instead of using light to illuminate specimens, the electron microscope employed a **beam of electrons**, which have a much shorter wavelength than visible light. This innovation allowed for magnifications far greater than those possible with traditional microscopes, revealing cellular structures at the nanometer scale. Two German physicists, **Ernst Ruska** and **Max Knoll**, played a pivotal role in developing the first practical electron microscope in 1931. Ruska's work, in particular, earned him the **Nobel Prize in Physics** in 1986, as it paved the way for imaging the previously invisible.

One of the most transformative applications of electron microscopy came in the study of **neurons**, the specialized cells that make up the nervous system. By the 1940s and 1950s, scientists used electron microscopes to capture detailed images of neurons, including their synapses, axons, and dendrites—features that were previously too small to visualize. This breakthrough allowed researchers to study the physical structure of neural networks and understand how neurons communicated with one another. For the first time, scientists could see the machinery of thought and sensation, laying the foundation for modern neuroscience.

Importance and Impact

Electron microscopy revolutionized the biological sciences by providing an **unprecedented view of cellular structures**. In the field of neuroscience, it revealed the physical basis of neural communication, enabling scientists to study the synaptic connections that form the foundation of the nervous system. This new level of detail allowed researchers to trace the pathways of electrical signals in the brain, helping to explain how information is processed, stored, and transmitted.

The ability to visualize neurons also had profound implications for **medical research**. Electron microscopy helped scientists understand the structural changes associated with neurological diseases, such as Alzheimer's, Parkinson's, and multiple sclerosis. By comparing healthy and diseased neural tissue, researchers gained critical insights into the cellular mechanisms underlying these conditions, paving the way for the development of treatments and interventions.

Beyond neuroscience, electron microscopy transformed **cell biology** by revealing the ultrastructure of organelles like mitochondria, the Golgi apparatus, and the endoplasmic reticulum. These high-resolution images helped elucidate the functions of these organelles, advancing our understanding of cellular processes such as energy production, protein synthesis, and intracellular transport. The tool also played a key role in discovering **viruses**, which were too small to be seen with optical microscopes, making electron microscopy essential for virology and the study of infectious diseases.

The applications of electron microscopy extended beyond biology. Material scientists used it to analyze the **atomic structure of materials**, leading to breakthroughs in metallurgy, nanotechnology, and semiconductor development. The ability to study materials at such a fine scale allowed for innovations in everything from computer chips to aerospace engineering. Electron microscopy became a cornerstone of **interdisciplinary research**, bridging the gap between biology, chemistry, physics, and engineering.

The advent of electron microscopy also highlighted the role of **technology in expanding the limits of human knowledge**. It demonstrated how new tools could uncover hidden worlds, sparking curiosity and further innovation. The images captured by electron microscopes were not just scientific data—they were awe-inspiring glimpses into the complexity of life and matter, providing both practical insights and philosophical questions about the nature of existence.

Chapter Summary

The development of electron microscopy in the 1930s opened a new era in science, allowing researchers to visualize structures at the nanometer scale. Its application to the study of neurons revolutionized neuroscience, providing insights into the physical basis of thought and sensation. Electron microscopy also transformed cell biology, virology, and material science, enabling discoveries that ranged from cellular organelles to the atomic structure of materials. By revealing the hidden architecture of life and matter, electron microscopy became one of the most important scientific tools of the 20th century, bridging disciplines and inspiring further exploration.

Next Chapter

As electron microscopy revealed the intricacies of cellular structures, other discoveries were making their way into everyday life. In the next chapter, we'll turn to an invention born of failure: **Bubble Wrap**, a material that was originally intended to be wallpaper but became an iconic packaging product. This story highlights how innovation often takes unexpected turns, leading to creations that redefine industries and solve problems in surprising ways.

027 Bubble Wrap Blunder
From Failed Wallpaper to Packing Revolution

The mid-20th century was an era of creative experimentation, where inventors sought to develop new materials and technologies to improve everyday life. One such pursuit began in 1957 when **Alfred Fielding**, a mechanical engineer, and **Marc Chavannes**, a Swiss chemist, set out to create a **textured wallpaper** that would appeal to modern design tastes. Their idea was to produce a wallpaper with a three-dimensional texture by sealing two sheets of plastic together with air pockets in between. Using a rudimentary heat-sealing machine, they bonded plastic film and accidentally created a sheet of **bubbled plastic**. While the result was not suitable for wallpaper, the unique material caught their attention.

Despite its failure as wallpaper, Fielding and Chavannes believed their bubbled plastic might have other applications. They initially tried marketing it as a **greenhouse insulation material**, reasoning that its air-filled pockets could retain heat while allowing light to pass through. However, this too failed to gain traction, as farmers and horticulturalists found it less effective and more expensive than traditional methods. Still, the inventors were determined to find a practical use for their unusual creation. They realized they had inadvertently invented a new type of material, but its purpose remained unclear.

The breakthrough came in 1960 when **IBM** launched its first mass-produced computer, the IBM 1401. The computer required careful handling during shipping due to its fragile components, and Fielding and Chavannes saw an opportunity. They repurposed their bubbled plastic as a **protective packaging material**, providing a lightweight, flexible cushion that could absorb shocks during transportation. The product, now known as **Bubble Wrap**, quickly became an essential tool in the shipping and packaging industries. Fielding and Chavannes founded **Sealed Air Corporation** to manufacture and market Bubble Wrap, and their accidental invention transformed the way goods were packaged and shipped worldwide.

Importance and Impact

Bubble Wrap revolutionized the packaging industry by offering a solution that was not only effective but also cost-efficient. Before Bubble Wrap, packaging fragile items often relied on heavy materials like crumpled paper, cardboard, or even sawdust, which were less effective at absorbing shocks and more labor-intensive to handle. Bubble Wrap's lightweight, flexible design made it easier to use and transport while providing superior protection. This innovation allowed businesses to ship delicate items more safely, reducing damage and saving costs in the process.

The invention of Bubble Wrap also supported the growth of industries that relied on the safe transport of fragile goods. For companies like IBM, whose success depended on delivering delicate computer components intact, Bubble Wrap became an indispensable tool. As the use of electronics grew, so too did the demand for protective packaging, with Bubble Wrap leading the way. Its success demonstrated how a product born of persistence and experimentation could create an entirely new market.

Beyond its practical uses, Bubble Wrap had cultural significance as well. It became synonymous with protection and care, symbolizing the safe handling of valuable items. Over time, it also gained unexpected popularity as a **stress-relief tool**, with people finding satisfaction in popping its air-filled bubbles. This phenomenon made Bubble Wrap a quirky cultural icon, showing how an accidental invention could resonate with people in ways its creators never intended.

Fielding and Chavannes' persistence in finding a use for their creation is a testament to the value of **creative problem-solving** and **adaptability** in innovation. Their story underscores an important lesson: not all inventions start with a clear purpose, but with determination, even failed ideas can evolve into transformative solutions. Bubble Wrap exemplifies how accidental discoveries can lead to success when inventors are willing to think beyond their initial goals.

Today, Bubble Wrap remains a staple of the packaging industry, with variations tailored to specific needs, such as anti-static versions for electronics and biodegradable options for environmentally conscious consumers. Its legacy reflects the ongoing impact of ingenuity and the importance of embracing unexpected outcomes in the creative process.

Chapter Summary

Bubble Wrap began as a failed attempt to create textured wallpaper but found its purpose as a revolutionary packaging material. Alfred Fielding and Marc Chavannes' persistence in repurposing their invention transformed it into a global success, providing an effective, lightweight solution for protecting fragile goods. Bubble Wrap not only revolutionized the shipping industry but also became a cultural phenomenon, valued for its practicality and its unexpected appeal as a stress-relief tool. This story highlights the power of adaptability in innovation, proving that even failed experiments can lead to breakthroughs with lasting impact.

Next Chapter

While Bubble Wrap addressed the challenges of packaging and shipping, other inventors were solving entirely different problems through similar accidents. In the next chapter, we'll explore the story of **Super Glue**, a failed attempt to create optical materials that instead resulted in one of the world's strongest adhesives. This accidental invention would revolutionize bonding materials across industries, from household repairs to medical applications.

028 Super Glue's Sticky Situation
A Failed Adhesive that Never Let Go

The invention of **Super Glue**, or **cyanoacrylate**, is a story of persistence and unintended success. In 1942, during World War II, **Dr. Harry Coover**, a chemist working for Eastman Kodak, was attempting to develop clear, heat-resistant plastics for use in precision gun sights. While experimenting with a new class of synthetic materials called **acrylates**, Coover and his team stumbled upon a compound that was incredibly sticky. However, the adhesive properties were seen as a problem, as the substance bonded instantly to virtually anything it touched. Frustrated by this unexpected stickiness, Coover abandoned the project, shelving the discovery as a failed attempt.

Years later, in 1951, Coover was overseeing another project at Eastman Kodak. This time, his team was developing heat-resistant polymers for jet canopies. Once again, they encountered the same sticky substance. Unlike before, Coover recognized the potential of this compound—not as a plastic for industrial use, but as a **powerful adhesive**. His insight came from its unique ability to form instant, durable bonds without requiring heat, pressure, or curing agents. He realized that this material could revolutionize how objects were joined together.

In 1958, Eastman Kodak patented the adhesive under the name **Eastman 910**, later rebranded as **Super Glue**. The product quickly gained popularity for its versatility, ease of use, and incredible strength. It was marketed to consumers for household repairs and to industries ranging from manufacturing to medicine. The success of Super Glue was a testament to Coover's ability to see value in an invention initially considered a failure, transforming it into a product that would become a staple of modern life.

Importance and Impact

Super Glue's impact extended far beyond its intended use. Its ability to create **instant, strong bonds** revolutionized the way people approached repairs and manufacturing. Unlike traditional adhesives, which required time to dry or external factors like heat, Super Glue provided a quick and effective solution for bonding materials ranging from wood and metal to glass and ceramics. This made it indispensable in industries where speed and reliability were essential, from automotive assembly to aerospace engineering.

In medicine, Super Glue found unexpected applications that saved lives. During the Vietnam War, medics used cyanoacrylate to seal wounds on the battlefield, temporarily stopping bleeding and stabilizing patients until they could receive proper medical treatment. This use inspired the development of medical-grade adhesives, which are now commonly used in surgeries to close incisions, reduce scarring, and eliminate the need for stitches in certain procedures. Super Glue's medical applications underscore its versatility and the transformative potential of a seemingly simple invention.

The adhesive also became an essential tool in forensic science. Investigators discovered that the vapors from cyanoacrylate could adhere to the ridges of fingerprints, creating clear, detailed impressions on surfaces where prints might otherwise be invisible. This technique, known as **cyanoacrylate fuming**, remains a cornerstone of modern fingerprint analysis and has helped solve countless criminal cases worldwide.

Super Glue's success illustrates how persistence and adaptability can transform a failed experiment into a groundbreaking invention. Dr. Coover's willingness to revisit his earlier "failure" and see its potential led to a product that transcended its original purpose. His work demonstrates the importance of keeping an open mind and recognizing opportunities in unexpected outcomes. Super Glue became more than an adhesive—it became a symbol of ingenuity, inspiring new ways to approach challenges in fields as diverse as medicine, manufacturing, and law enforcement.

Today, Super Glue is a household name and continues to be a vital tool in industries and homes alike. Variants have been developed to suit specific needs, including waterproof and flexible formulas. Its legacy lies in its simplicity and effectiveness, proving that even the most accidental discoveries can leave a lasting impact on the world.

Chapter Summary

Super Glue's invention was an accidental triumph that emerged from failed experiments in developing heat-resistant materials. Dr. Harry Coover's ability to see potential in cyanoacrylate's unique adhesive properties transformed it into a versatile product used in homes, industries, and even battlefields. Its applications range from simple household repairs to life-saving medical treatments and forensic investigations, illustrating the far-reaching impact of this unexpected discovery. Super Glue's story is a testament to the value of persistence, adaptability, and the ability to turn failure into innovation.

Next Chapter

As Super Glue found unexpected applications across various fields, other innovations born from necessity were also reshaping medicine. In the next chapter, we'll explore the origins of **plastic surgery** during World War I, where battlefield conditions drove surgeons to develop pioneering reconstructive techniques. These early efforts not only saved lives but also laid the foundation for modern plastic and reconstructive surgery.

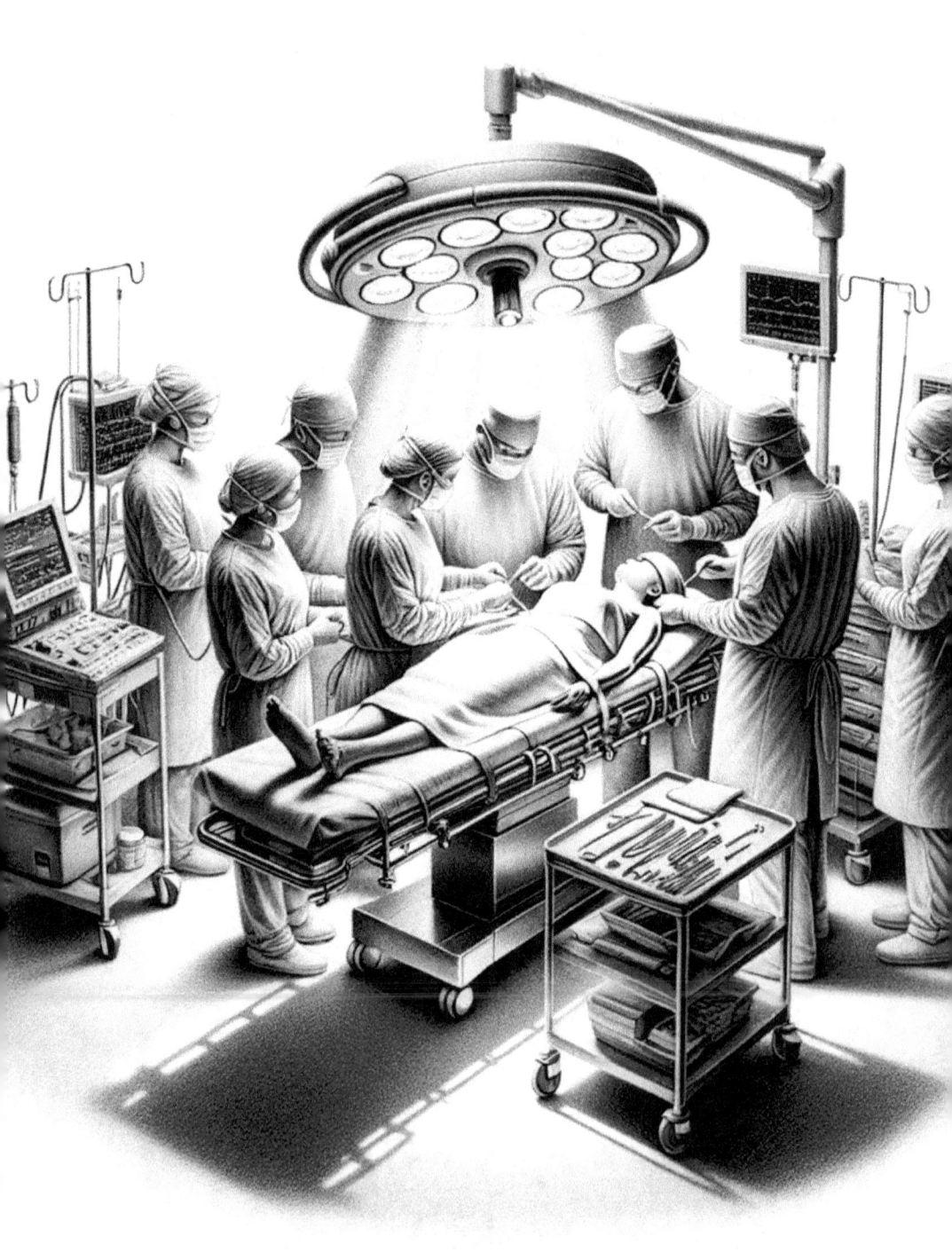

029 Plastic Surgery's Battle Origins

World War I Necessities and Modern Reconstructive Surgery

The brutality of **World War I** brought unprecedented challenges to the field of medicine. For the first time, modern weaponry—artillery, shrapnel, and chemical warfare—inflicted devastating injuries that disfigured countless soldiers. Unlike previous conflicts, where such injuries often led to death, advances in medical care, such as improved anesthesia and antiseptics, enabled more soldiers to survive. However, many survivors returned home with **severe facial and bodily disfigurements**, creating an urgent need for reconstructive techniques that had never been systematically explored.

Before the war, plastic surgery was a fledgling field, with only a handful of surgeons possessing the skills and creativity to attempt **reconstructive procedures**. Among them was **Harold Gillies**, a New Zealand-born surgeon working in Britain. Gillies was inspired by his colleague **Charles Valadier**, a French-American dentist who had experimented with facial reconstructions using dental materials. In 1917, Gillies established a specialized hospital for facial injuries in **Sidcup**, England, where he pioneered techniques to repair the shattered faces of soldiers. His work was not just medical but deeply personal; Gillies recognized the profound psychological impact of disfigurement and sought to restore dignity and humanity to his patients.

Gillies and his team developed groundbreaking methods, such as **tubed pedicle grafts**, which involved transferring skin from one part of the body to another while keeping it attached to its original blood supply. This technique minimized the risk of infection and revolutionized the possibilities for reconstructive surgery. Gillies treated more than **5,000 soldiers** during the war, creating a foundation for the modern practice of **plastic surgery**. His meticulous documentation and photographs of his procedures ensured that his techniques would influence future generations of surgeons.

Importance and Impact

The innovations born out of necessity during World War I laid the groundwork for **modern reconstructive and cosmetic surgery**. Gillies' techniques were not only functional but also focused on aesthetics, emphasizing the importance of both physical and psychological recovery. His work demonstrated that reconstructive surgery could restore not just physical appearance but also confidence and identity, addressing the social stigma faced by disfigured veterans.

The impact of these early techniques extended far beyond the battlefield. During the interwar period, Gillies' methods were adapted for civilian use, addressing injuries from industrial accidents, burns, and congenital conditions. His work inspired other surgeons, such as his cousin **Archibald McIndoe**, who would later build upon Gillies' methods during World War II to treat pilots with severe burns. These advancements in reconstructive surgery continued to evolve, eventually leading to modern procedures like **microsurgery**, which allows surgeons to reconnect tiny blood vessels and nerves.

Gillies' work also laid the ethical foundation for plastic surgery as a legitimate medical field. Before his contributions, reconstructive surgery was often seen as experimental or even frivolous. His dedication to documentation and teaching ensured that his techniques were standardized and widely disseminated, transforming the field from a niche practice into a respected branch of surgery. Today, reconstructive surgery is an integral part of modern medicine, encompassing everything from trauma repair to life-changing procedures for cancer survivors.

The psychological impact of reconstructive surgery cannot be overstated. For the soldiers Gillies treated, regaining their appearance meant reclaiming a sense of normalcy and identity. His work highlighted the intersection of medicine and mental health, demonstrating that healing was not only about physical repair but also about restoring a person's place in society. This holistic approach remains a cornerstone of modern plastic surgery, where the goal is to address the **whole person**—body and mind.

The innovations of World War I also planted the seeds for the rise of **cosmetic surgery**. Techniques developed for reconstructive purposes began to be adapted for aesthetic enhancements. While initially controversial, cosmetic surgery grew in popularity, expanding the scope of plastic surgery and opening new possibilities for personal expression and self-confidence. This evolution of the field underscores the transformative power of medical innovation driven by necessity.

Chapter Summary

The challenges of World War I forced surgeons like Harold Gillies to invent groundbreaking techniques that laid the foundation for modern plastic and reconstructive surgery. From repairing the disfigured faces of soldiers to addressing the psychological toll of their injuries, these innovations transformed a fledgling field into a critical branch of medicine. Gillies' legacy continues to shape the practice of reconstructive surgery, emphasizing both physical and emotional recovery. The advancements made during this time would not only help soldiers reclaim their identities but also inspire civilian applications, paving the way for modern reconstructive and cosmetic surgery.

Next Chapter

As surgeons addressed the challenges of reconstructing human faces, scientists were simultaneously uncovering unseen patterns within the human body. In the next chapter, we'll explore the development of **Magnetic Resonance Imaging (MRI)**, a technology that transformed medical diagnostics by allowing doctors to see inside the body with unprecedented clarity. This breakthrough revealed the invisible structures of our anatomy, opening a new frontier in medical imaging.

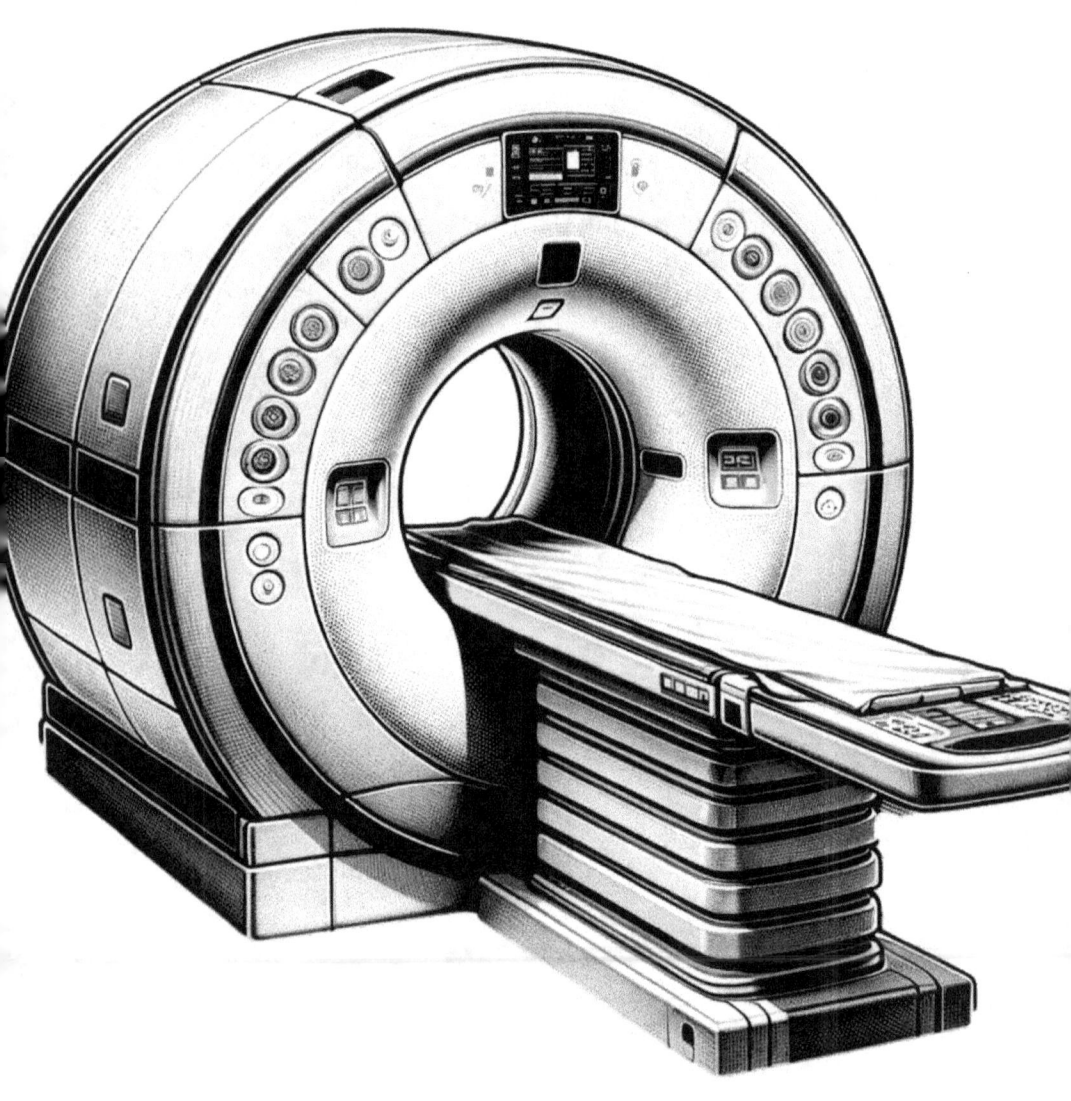

030 Magnetic Resonance
Unseen Patterns Become MRI Scans

The journey to developing **Magnetic Resonance Imaging (MRI)** began in the early 20th century, rooted in the fields of **physics** and **chemistry** rather than medicine. In 1938, physicist **Isidor Isaac Rabi** discovered a phenomenon he called **nuclear magnetic resonance (NMR)**, which revealed that certain atomic nuclei absorbed and re-emitted electromagnetic energy when exposed to a magnetic field. This finding earned Rabi the **1944 Nobel Prize in Physics** and laid the groundwork for understanding how magnetic fields interact with matter at the atomic level.

By the 1940s, scientists like **Felix Bloch** and **Edward Purcell** further refined the NMR process, enabling the detection of specific atomic structures. Their work earned them the **1952 Nobel Prize in Physics** and paved the way for using NMR in analytical chemistry to study the molecular composition of substances. However, applying these principles to the human body would require significant technological and conceptual leaps. It wasn't until the 1970s that NMR evolved into a medical imaging tool, thanks to the efforts of researchers like **Raymond Damadian**, who theorized that the differences in magnetic properties between cancerous and normal tissues could be used to diagnose disease.

Damadian, along with physicists **Paul Lauterbur** and **Peter Mansfield**, spearheaded the transformation of NMR into what is now known as MRI. Lauterbur introduced the idea of using magnetic gradients to create **two-dimensional images**, and Mansfield developed mathematical techniques to improve image resolution and speed. Their breakthroughs enabled MRI to produce detailed, non-invasive images of internal organs and tissues, revolutionizing diagnostics. By the early 1980s, MRI machines became a fixture in hospitals, offering doctors an unprecedented view inside the human body.

Importance and Impact

The invention of MRI fundamentally changed the way diseases are diagnosed and treated. Unlike X-rays or CT scans, which rely on ionizing radiation, MRI uses harmless magnetic fields and radio waves to create highly detailed images of soft tissues. This made it particularly effective for examining organs like the brain, heart, and spinal cord, as well as detecting abnormalities such as tumors, strokes, and injuries. MRI's ability to provide clear, **three-dimensional** views of internal structures revolutionized fields like **neurology**, **orthopedics**, and **oncology**, offering insights that were previously unattainable.

MRI also became an indispensable tool in **research**, enabling scientists to study the human body in real-time. Techniques like **functional MRI (fMRI)**, which measures brain activity by detecting changes in blood flow, have opened new frontiers in neuroscience. Researchers now use fMRI to map brain functions, study mental health conditions, and explore the neural basis of human behavior. This has deepened our understanding of conditions such as Alzheimer's disease, depression, and autism, as well as the brain's response to stimuli like music, language, and pain.

Beyond its diagnostic applications, MRI has had a profound societal impact by improving patient outcomes and reducing the need for invasive procedures. Before MRI, exploratory surgeries were often required to diagnose internal issues, posing significant risks to patients. MRI eliminated much of this uncertainty, allowing for earlier and more accurate diagnoses, which in turn improved treatment options and survival rates. The technology's non-invasive nature also made it more accessible and safer for patients, further enhancing its utility in medical care.

The collaborative nature of MRI's development highlights the importance of interdisciplinary research. It brought together physicists, chemists, engineers, and physicians to solve complex problems, demonstrating the power of teamwork in scientific innovation. MRI's success also underscored the role of basic science research in driving technological advancements. What began as an exploration of atomic behavior ultimately became one of the most important medical tools of the modern era.

Today, MRI continues to evolve, with advancements in speed, resolution, and specialized imaging techniques such as **diffusion tensor imaging (DTI)**, which maps the brain's white matter pathways. These innovations ensure that MRI remains at the forefront of medical diagnostics, pushing the boundaries of what we can see and understand about the human body.

Chapter Summary

Magnetic Resonance Imaging (MRI) emerged from decades of research into nuclear magnetic resonance, evolving into one of the most transformative tools in modern medicine. By providing non-invasive, detailed images of internal organs and tissues, MRI revolutionized diagnostics and treatment, particularly in fields like neurology and oncology. The invention of MRI was a collaborative effort that bridged physics, chemistry, and medicine, showcasing the power of interdisciplinary research. From improving patient care to advancing scientific understanding, MRI remains a cornerstone of modern healthcare, continually expanding its capabilities and impact.

Next Chapter

As MRI transformed medical diagnostics, another innovation was reshaping the way scientists manipulated light. The next chapter delves into the invention of the **laser**, a technology born from theoretical physics that would revolutionize everything from communications to medicine. We'll explore how the laser turned abstract ideas into a beam of coherent light that became a cornerstone of modern technology.

031 The Laser Leap
Theory Turned Into Blazing Light

The invention of the **laser** is a story that began in the realm of theoretical physics and culminated in one of the most versatile and revolutionary technologies of the 20th century. The journey started with **Albert Einstein**, who, in 1917, introduced the concept of **stimulated emission of radiation** in his work on quantum theory. This principle suggested that under certain conditions, atoms could emit light in a way that would amplify it, creating a highly focused beam. While this idea remained purely theoretical for decades, it inspired a generation of physicists to explore the practical possibilities of controlled light amplification.

The path to the **laser** (an acronym for **Light Amplification by Stimulated Emission of Radiation**) began to take shape in the mid-20th century. In 1953, **Charles Townes** and his colleagues developed the **maser (Microwave Amplification by Stimulated Emission of Radiation)**, which operated on similar principles but amplified microwaves instead of visible light. The maser demonstrated that Einstein's theory could be applied to generate coherent radiation, but the leap to visible light required further breakthroughs. In 1960, physicist **Theodore Maiman** at Hughes Research Laboratories successfully built the first laser, using a synthetic ruby crystal to produce a narrow, intense beam of red light. Maiman's laser was not only a proof of concept but also the beginning of a new era in optics and technology.

At first, the laser was often described as "a solution in search of a problem." Its practical applications were not immediately obvious, and early demonstrations of its power often seemed more like science fiction than science. However, as scientists and engineers began to experiment with the technology, they discovered that lasers could be used in an astonishing range of fields, from communications to medicine to entertainment. What started as an abstract concept rooted in quantum mechanics soon became one of the most important and ubiquitous tools in modern life.

Importance and Impact

The laser's ability to produce a focused, coherent beam of light with specific wavelengths made it a **transformational technology**. One of its earliest and most important applications was in **telecommuni-

cations, where lasers became the backbone of **fiber-optic networks**, enabling the rapid transmission of data over long distances. By encoding information into light signals, lasers revolutionized how data was transmitted, laying the foundation for the **internet** and modern global connectivity. Today, the laser's role in telecommunications remains indispensable, powering everything from high-speed internet to satellite communications.

In medicine, the laser has had a profound impact, offering new ways to treat conditions with precision and minimal invasiveness. Laser surgery, used in procedures ranging from eye corrections to tumor removal, allows doctors to perform delicate operations with unprecedented accuracy. The laser's precision has also been harnessed in **diagnostics**, such as in the development of laser-based imaging techniques like **confocal microscopy**, which provides detailed views of biological tissues. Beyond surgery and imaging, lasers are used in dermatology, dentistry, and even in non-invasive cancer treatments, showcasing their versatility in improving healthcare outcomes.

The laser's influence extends to fields like **manufacturing**, where it is used for cutting, welding, and engraving materials with high precision. In **research**, lasers have opened new frontiers in physics, chemistry, and biology. Techniques like **laser cooling** have advanced our understanding of quantum mechanics, while laser-based spectroscopy has enabled scientists to study the molecular composition of distant planets. Lasers have also been integral to the development of **optical tweezers**, which manipulate microscopic particles with light, earning Arthur Ashkin, Gérard Mourou, and Donna Strickland the 2018 Nobel Prize in Physics..

Culturally, the laser has left its mark on entertainment and art, from dazzling light shows to the precision of optical media like **CDs** and **DVDs**. The creation of holograms, another application of laser technology, has become a staple of futuristic displays and even archival preservation. The laser's versatility and cultural resonance underscore its status as a true technological revolution.

The development of the laser highlights the importance of **fundamental research** and the unpredictable pathways of innovation. What began as a theoretical insight by Einstein became a practical tool through the work of scientists like Townes and Maiman, and its applications far exceeded their initial expectations. The laser serves as a reminder that the most profound technological advancements often arise from curiosity-driven exploration and a willingness to see beyond the immediate utility of an idea.

Chapter Summary

The laser, born from Einstein's theory of stimulated emission and realized through the efforts of pioneers like Theodore Maiman, transformed the landscape of modern technology. Its applications span fields as diverse as telecommunications, medicine, manufacturing, and entertainment, making it one of the most versatile inventions of the 20th century. From fiber-optic networks to life-saving surgeries and groundbreaking research, the laser's precision and power continue to shape the world. The story of its development underscores the value of fundamental research and the boundless potential of human ingenuity.

Next Chapter

While the laser revolutionized communication and industry, another group of innovators was building the foundation for what would become the tech epicenter of the world. In the next chapter, we'll explore the origins of **Silicon Valley**, focusing on the bold decisions of the **Fairchild Eight** and the creation of Fairchild Semiconductor, which catalyzed the rise of a technology hub that would change the course of history.

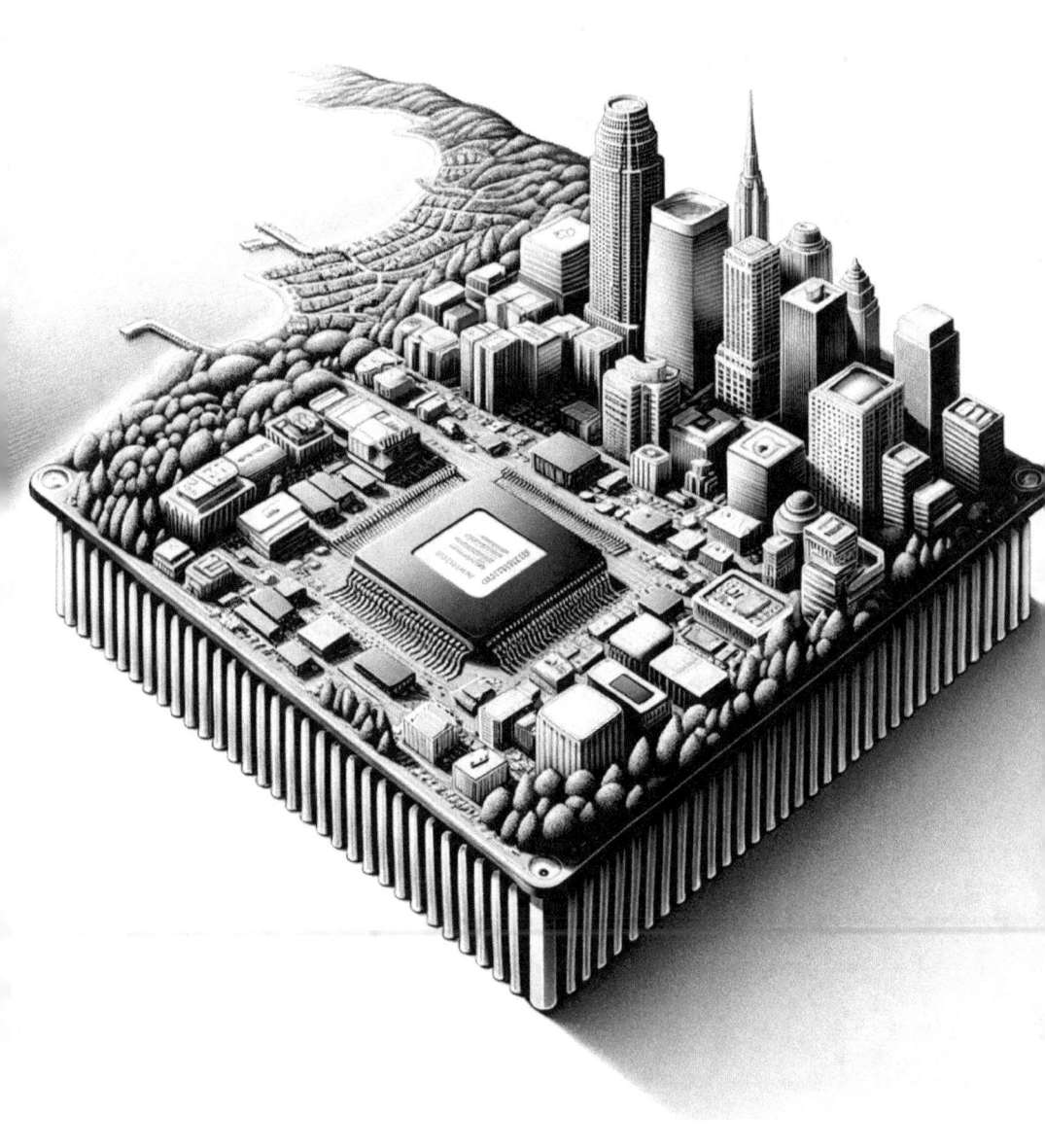

032 Silicon Valley's Seeds
Fairchild Semiconductor and Silicon's Future

The roots of **Silicon Valley**, the global hub of technology and innovation, can be traced back to the late 1950s. At the center of this transformation was a small group of engineers who would come to be known as the **Fairchild Eight**. Their story begins with **William Shockley**, co-inventor of the transistor and Nobel Prize laureate, who established **Shockley Semiconductor Laboratory** in Mountain View, California, in 1956. Shockley's vision was to develop cutting-edge transistor technology using silicon, a material that promised greater efficiency than the germanium transistors commonly used at the time. To achieve this, he recruited some of the brightest young engineers, including **Gordon Moore**, **Robert Noyce**, and six others.

Despite Shockley's technical brilliance, his erratic leadership and authoritarian management style created tension within his team. The engineers grew increasingly frustrated with his approach, which stifled collaboration and innovation. In 1957, the group made the bold decision to leave Shockley Semiconductor and form their own company. Backed by funding from **Fairchild Camera and Instrument Corporation**, they founded **Fairchild Semiconductor**, setting up shop just a few miles away.

Fairchild Semiconductor quickly became a pioneer in the burgeoning semiconductor industry. The company developed the first commercially viable **silicon transistors**, which were smaller, faster, and more reliable than their germanium predecessors. These innovations not only established Fairchild as a leader in the field but also laid the foundation for the **integrated circuit**, a groundbreaking technology that would revolutionize computing and electronics.

Importance and Impact

The creation of Fairchild Semiconductor was a defining moment in the history of technology, marking the birth of what would later be called **Silicon Valley**. The success of the Fairchild Eight demonstrated that **risk-taking**, **innovation**, and **entrepreneurship** could thrive in a supportive environment. This culture of openness and collaboration became a hallmark of Silicon Valley, attracting talent and investment from around the world.

Fairchild's innovations in silicon transistors and manufacturing techniques were critical to the evolution of the **semiconductor industry**, which forms the backbone of modern electronics. The company's contributions enabled the development of smaller and more powerful devices, from computers to mobile phones. The use of silicon, abundant and cost-effective, transformed transistors from a niche technology into the cornerstone of the digital revolution.

Equally important was Fairchild Semiconductor's role in fostering a generation of **technology entrepreneurs**. Many of its key figures, including Gordon Moore and Robert Noyce, went on to found **Intel**, which would dominate the semiconductor market and drive the development of microprocessors. The "Fairchild Effect" spread throughout Silicon Valley as former employees started their own companies, creating a network of startups that propelled the region into a global technology hub.

The culture of innovation at Fairchild also inspired new business practices, including the emphasis on **venture capital** as a means to fund risky but high-potential ideas. Investors began to see the immense value in technology startups, leading to the establishment of a robust ecosystem that nurtured innovation. This shift in funding models further cemented Silicon Valley's position as a breeding ground for groundbreaking technologies.

Fairchild Semiconductor's legacy extends beyond the products it created. Its influence on the culture, structure, and approach of the technology industry has shaped how companies operate to this day. The emphasis on collaboration, agility, and a willingness to take risks remains a defining characteristic of Silicon Valley, ensuring its continued relevance in the global economy.

Chapter Summary

Fairchild Semiconductor, founded by the Fairchild Eight after their departure from Shockley Semiconductor, marked the beginning of Silicon Valley as the world's premier technology hub. By pioneering silicon transistors and fostering a culture of innovation, Fairchild transformed the semiconductor industry and set the stage for the digital revolution. The company's success inspired a generation of entrepreneurs, creating a network of startups that defined Silicon Valley's ecosystem. Fairchild's legacy lies not only in its technological achievements but also in its influence on the culture of risk-taking and collaboration that continues to drive technological progress.

Next Chapter

Building on the innovations of Fairchild Semiconductor, the next major breakthrough in electronics came with the development of the **integrated circuit**. In the following chapter, we'll explore how this revolutionary invention, born out of the challenges of miniaturizing electronics, paved the way for modern computing and became the foundation of nearly all digital technology.

033 The Integrated Circuit

A Tiny Revolution in Computing

By the late 1950s, the field of electronics faced a critical challenge: as devices grew more complex, so did the number of individual components, such as transistors, resistors, and capacitors, required to build them. This led to the problem of **"the tyranny of numbers"**—the physical and logistical difficulty of wiring thousands of tiny components together reliably. Engineers realized that further miniaturization and increased complexity would require a fundamentally new approach to circuit design.

Two inventors working independently addressed this challenge in 1958: **Jack Kilby** at Texas Instruments and **Robert Noyce** at Fairchild Semiconductor. Kilby, a newly hired engineer, created the **first prototype of an integrated circuit (IC)** using a single piece of germanium to house multiple electronic components. His circuit was a simple oscillator, but it demonstrated that components could be fabricated together as a single unit. Meanwhile, Noyce developed a competing design using silicon and a process called **planar technology**, which allowed circuits to be etched directly onto the silicon surface. Noyce's approach proved more practical for mass production and eventually became the industry standard.

Kilby's and Noyce's breakthroughs solved the problem of interconnecting thousands of components by integrating them onto a single chip, vastly reducing the size, weight, and cost of electronic systems. This innovation marked the beginning of the **microelectronics revolution**, laying the foundation for modern computing and electronics.

Importance and Impact

The integrated circuit transformed the world of electronics, enabling the creation of smaller, faster, and more efficient devices. Early applications included military systems such as missile guidance and communications, where size and reliability were critical. However, it wasn't long before ICs found their way into commercial products, such as calculators and hearing aids, demonstrating their versatility and potential for mass-market adoption.

One of the most significant impacts of the integrated circuit was its role in the development of the **microprocessor**. By combining multiple

ICs onto a single chip, engineers created the first **general-purpose processors**, which could perform a wide variety of tasks. This breakthrough fueled the rise of **personal computers**, **gaming consoles**, and **mobile devices**, fundamentally changing how people work, play, and communicate.

The integrated circuit also revolutionized manufacturing processes, creating an entirely new industry centered on **semiconductor fabrication**. This industry drove rapid technological advances, following **Moore's Law**, which predicted that the number of transistors on a chip would double approximately every two years. The relentless pace of innovation in IC technology led to exponential growth in computing power, making possible everything from smartphones to artificial intelligence.

Beyond its technological impact, the integrated circuit had profound economic and cultural effects. It democratized access to technology, making powerful tools available to individuals and businesses alike. By reducing costs and increasing reliability, ICs paved the way for the digital age, in which computing and connectivity are integral to everyday life. The rise of consumer electronics, the internet, and global communications can all be traced back to this transformative invention.

The story of the integrated circuit is also one of collaboration and competition. Kilby and Noyce's parallel discoveries highlight the importance of independent yet complementary approaches to problem-solving. Their innovations not only solved immediate technical challenges but also established a model for **interdisciplinary collaboration**, bringing together physics, engineering, and manufacturing expertise to create entirely new possibilities.

Chapter Summary

The invention of the integrated circuit by Jack Kilby and Robert Noyce revolutionized electronics by integrating multiple components onto a single chip, solving the problem of complexity and enabling unprecedented miniaturization. This breakthrough laid the foundation for modern computing, making possible everything from microprocessors to personal devices. The integrated circuit transformed not only technology but also society, ushering in the digital age and reshaping how people live, work, and communicate. Its legacy continues to drive innovation, powering advancements in computing, communications, and beyond.

Next Chapter

As the integrated circuit propelled electronics into the digital age, another technological breakthrough was lighting the way forward—literally. In the next chapter, we'll explore the story of the **light-emitting diode (LED)**, a small but mighty invention that revolutionized lighting and transformed industries ranging from displays to energy efficiency.

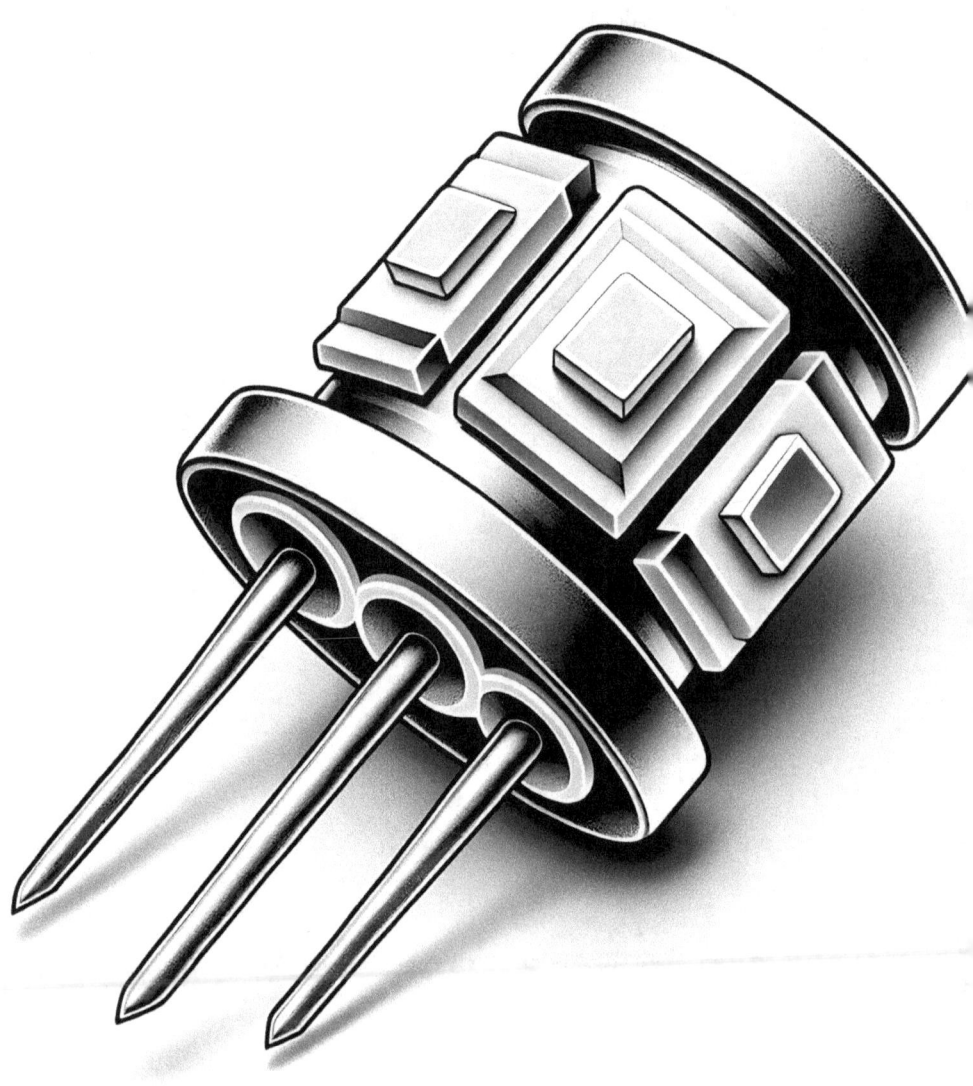

034 LED's First Light
A Small Glow Igniting Efficient Lighting

The invention of the **light-emitting diode (LED)** is a story of persistence, collaboration, and the exploration of fundamental physics. In the early 20th century, scientists began to observe that certain materials emitted light when an electric current passed through them, a phenomenon later known as **electroluminescence**. Early experiments in the 1920s and 1930s explored this effect in materials like silicon carbide, but these rudimentary devices produced dim light and had no practical applications.

The breakthrough came in 1962 when **Nick Holonyak Jr.**, an engineer at General Electric, developed the first practical LED capable of emitting visible red light. Holonyak, often called the "father of the LED," built on earlier work by scientists exploring semiconductors and the behavior of electrons in crystalline materials. By using a gallium arsenide phosphide (GaAsP) semiconductor, Holonyak's LED emitted a bright, stable red light when powered, making it the first step toward practical solid-state lighting.

Although the initial LEDs were primarily used as indicator lights in electronics, their potential as an efficient and durable light source was quickly recognized. Over the following decades, researchers refined LED technology, expanding its color range to include green, blue, and eventually white light. This evolution transformed LEDs from niche components into the cornerstone of modern lighting systems.

Importance and Impact

The invention of the LED was a turning point in lighting and energy efficiency. Unlike traditional incandescent bulbs, which generate light by heating a filament, LEDs produce light through the movement of electrons within a semiconductor. This process is far more energy-efficient, as it minimizes heat generation and converts most of the energy directly into light. Early LEDs were limited in brightness and color range, but advances in materials science and manufacturing enabled them to become powerful and versatile light sources.

One of the most significant applications of LEDs is in **energy-efficient lighting**, which has dramatically reduced global electricity consumption. Traditional lighting systems waste a significant portion of energy as heat, but LEDs consume a fraction of the power while providing the same or better illumination. This has had a profound impact on efforts to reduce greenhouse gas emissions and promote sustainable energy use.

LEDs have also revolutionized **display technology**. From the first calculator and digital watch displays to modern televisions, smartphones, and computer screens, LEDs have made vibrant, high-resolution displays possible. The invention of organic LEDs (OLEDs) further expanded their applications, enabling thinner, flexible screens used in cutting-edge devices.

In addition to energy and display technology, LEDs have been transformative in fields like **medicine**, **automotive engineering**, and **architecture**. Their compact size and versatility allow for innovations such as precision surgical lighting, adaptive headlights, and dynamic building designs. The ability to control LED color and brightness with high precision has also made them indispensable in art installations and theatrical productions.

The success of the LED highlights the value of **incremental innovation**, as its development spanned decades and involved contributions from multiple disciplines and industries. The pursuit of better materials, manufacturing processes, and applications has made LEDs a cornerstone of modern technology, demonstrating how persistent refinement can unlock the full potential of a seemingly simple invention.

Chapter Summary

The invention of the LED by Nick Holonyak Jr. in 1962 marked the beginning of a revolution in lighting and energy efficiency. From their humble origins as indicator lights, LEDs evolved into a powerful technology used in energy-efficient lighting, display systems, and numerous other applications. Their impact spans industries ranging from consumer electronics to medicine, reducing energy consumption and enabling new innovations. The story of the LED exemplifies how scientific curiosity and technological persistence can transform everyday life and drive sustainable progress.

Next Chapter

As LEDs began lighting the way forward, another innovation was emerging that would become the heart of modern computing: the **microprocessor**. In the next chapter, we'll explore how this tiny chip transformed technology by integrating the power of computing into an incredibly compact form, sparking the digital revolution that continues to shape our world today.

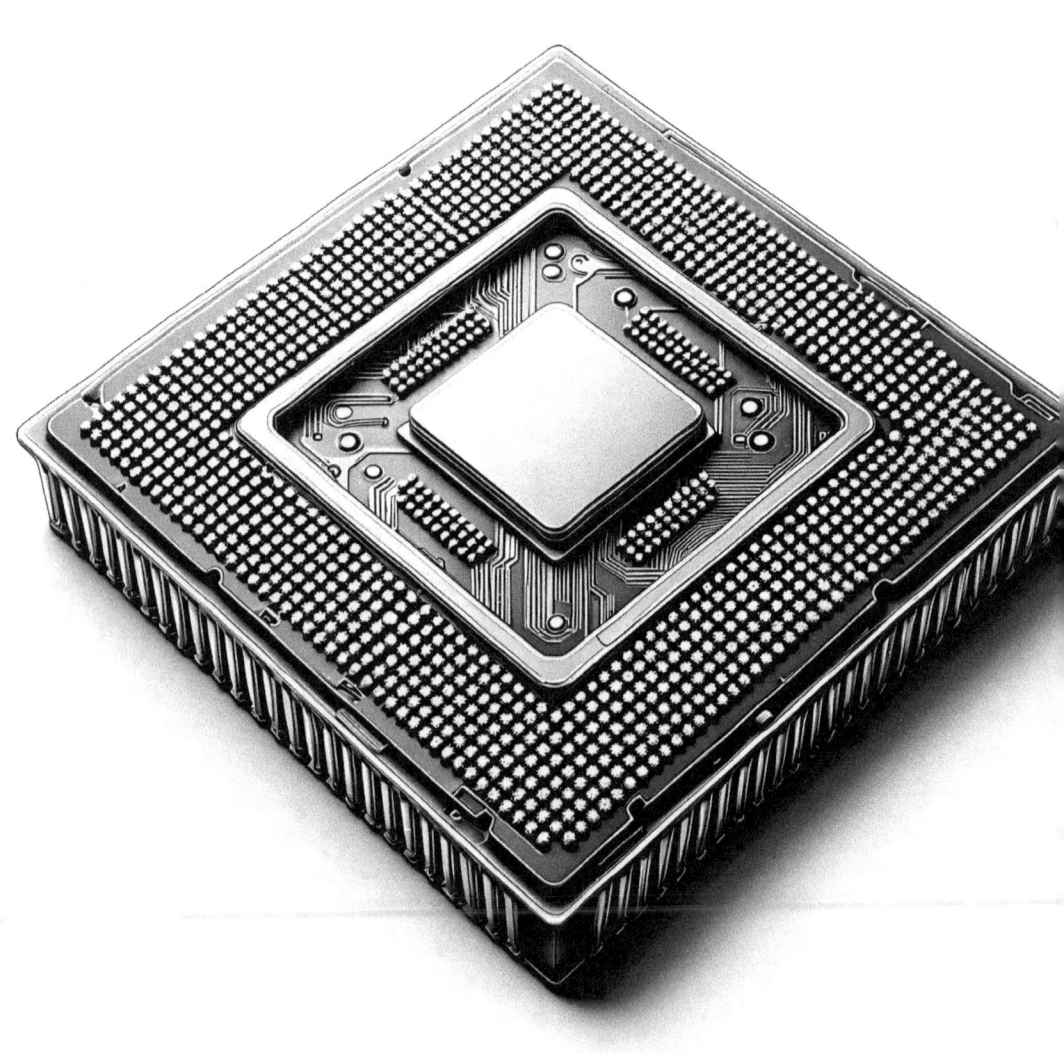

035 Microprocessor
How a Simple Chip Transformed Technology

By the early 1970s, the world of electronics was on the cusp of a transformative leap. The invention of the **integrated circuit (IC)** had already revolutionized the design of electronic systems, enabling engineers to shrink multiple components onto a single chip. However, building a complete computing system still required multiple ICs, interconnected in complex ways. The challenge of further miniaturizing these systems to fit into smaller devices required an entirely new innovation: the **microprocessor**.

The story of the microprocessor began in 1969, when the Japanese company **Busicom** approached the American firm **Intel** to design a set of specialized ICs for a new calculator. Instead of creating separate chips for each function, **Ted Hoff**, an engineer at Intel, proposed integrating all the calculator's processing functions onto a single chip. With the help of **Federico Faggin**, who developed the necessary silicon-gate technology, and **Stan Mazor**, who contributed to the architecture, Intel created the **4004 microprocessor**, released in 1971. This tiny chip contained **2,300 transistors** and performed up to **92,000 operations per second**, making it the first commercially available microprocessor.

Although initially intended for calculators, the microprocessor's potential quickly became evident. Its ability to execute programmable instructions meant it could serve as the "brain" of a wide range of electronic devices. This flexibility transformed the microprocessor into a cornerstone of modern computing, paving the way for the **personal computer revolution** and the integration of computing power into countless technologies.

Importance and Impact

The microprocessor fundamentally changed the way computers were designed, built, and used. Before its invention, computers were large, expensive machines used primarily by governments, universities, and large corporations. The microprocessor enabled the creation of **smaller**, **cheaper**, and **more accessible computers**, democratizing access to computing power. This shift led directly to the development of the **personal computer (PC)**, which would transform how individuals worked, learned, and communicated.

One of the earliest applications of the microprocessor was in the **Altair 8800**, a pioneering personal computer introduced in 1975. The Altair's success demonstrated the microprocessor's potential to bring computing into homes and small businesses, setting the stage for companies like Apple and Microsoft to capitalize on the growing demand for user-friendly PCs. As microprocessors grew more powerful, they enabled the development of increasingly sophisticated software, from word processors to video games, driving the rapid expansion of the technology industry.

Beyond personal computing, the microprocessor became the foundation for countless other technologies. In **consumer electronics**, it powered devices like digital watches, calculators, and video game consoles. In **automotive engineering**, it enabled the development of electronic control systems that improved fuel efficiency and safety. In **telecommunications**, microprocessors revolutionized network infrastructure and mobile devices, making global communication faster and more reliable.

The microprocessor also drove advances in **embedded systems**, where computing power is integrated into everyday objects. From household appliances to industrial machinery, embedded microprocessors have made technology more efficient, intelligent, and interconnected. This concept, which underpins the modern **Internet of Things (IoT)**, allows devices to communicate and function autonomously, transforming industries ranging from healthcare to agriculture.

The invention of the microprocessor also catalyzed a cultural shift, as computing became an integral part of daily life. By empowering individuals and small businesses with affordable computing power, the microprocessor enabled a wave of innovation that reshaped the global economy. From its humble beginnings as a calculator chip, the microprocessor evolved into a symbol of human ingenuity and a driver of progress in the digital age.

Chapter Summary

The microprocessor, pioneered by Intel's 4004 chip in 1971, revolutionized computing by integrating processing power into a single, compact device. This innovation democratized access to technology, enabling the development of personal computers and countless other applications. From consumer electronics to telecommunications and the Internet of Things, the microprocessor became the foundation of modern technology. Its impact extends far beyond engineering, transforming industries, reshaping society, and empowering individuals to harness the power of computing in their everyday lives.

Next Chapter

As microprocessors began powering a new wave of innovation, another revolutionary idea was emerging that would connect computers across vast distances. In the next chapter, we'll explore the origins of the **internet**, tracing its development from a military experiment to a global network that redefined communication and information sharing in the digital age.

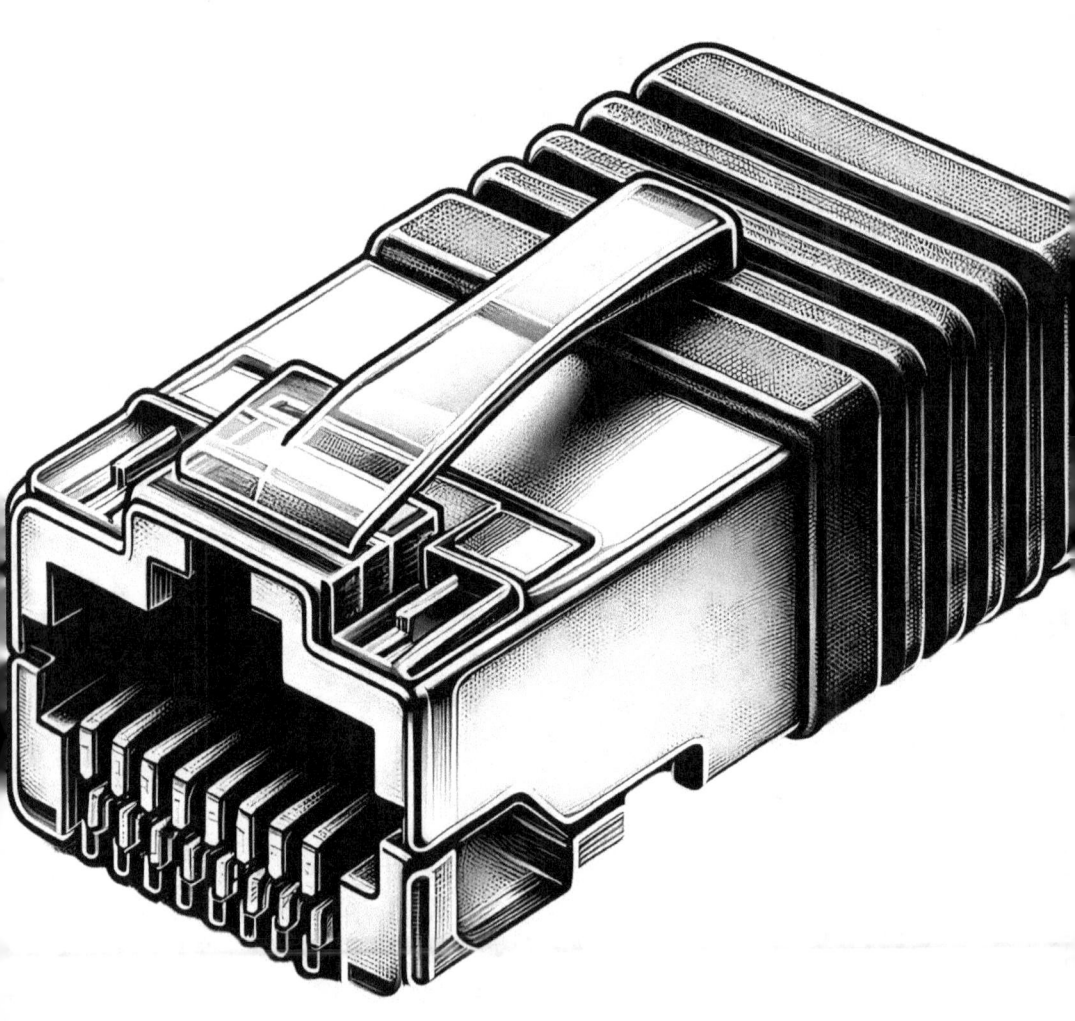

036 The Internet's First Nodes
Connecting Computers in ARPANET

The idea of creating a network to connect computers began as a response to the geopolitical tensions of the **Cold War**. In the late 1950s and early 1960s, the **Advanced Research Projects Agency (ARPA)**, a division of the U.S. Department of Defense, sought to develop technologies that could maintain reliable communication in the event of a nuclear attack. Traditional communication networks relied on centralized hubs, making them vulnerable to disruption. ARPA's vision was to create a **decentralized network**, where information could travel through multiple pathways, ensuring its delivery even if parts of the system were destroyed.

In 1969, this vision became a reality with the launch of **ARPANET**, the world's first operational packet-switching network. Developed by ARPA and a group of pioneering computer scientists, ARPANET allowed computers at geographically distant locations to communicate directly with one another. The network's first node was established at **UCLA - University of California**, followed shortly by nodes at **Stanford Research Institute, UC Santa Barbara**, and the **University of Utah**. On October 29, 1969, the first message—"LO"—was sent between UCLA and Stanford before the system crashed. Despite the technical hiccup, this marked the birth of what would eventually become the **internet**.

The key innovation behind ARPANET was **packet switching**, a method of breaking down information into small packets that could travel independently across the network and be reassembled at their destination. This approach, developed by researchers like **Paul Baran, Donald Davies**, and **Leonard Kleinrock**, allowed for more efficient and reliable data transmission than traditional circuit-switching methods. It also laid the groundwork for the development of **protocols** like TCP/IP, which would become the universal language of the internet.

Importance and Impact

The creation of ARPANET was a watershed moment in the history of communication, marking the transition from isolated computing systems to interconnected networks. While its initial purpose was to facilitate collaboration among researchers working on government-funded projects, its broader potential soon became apparent. ARPANET allowed scientists to **share resources**, **exchange data**, and **collab-**

orate in real-time, making it an indispensable tool for academic and technological innovation.

The decentralized nature of ARPANET also inspired new ways of thinking about communication and information sharing. It demonstrated that computers could serve not only as individual tools but as nodes in a larger system, capable of distributing and processing information collectively. This shift laid the intellectual foundation for concepts like **distributed computing**, which underpin modern technologies from cloud computing to blockchain.

ARPANET also served as a testing ground for technologies that would become integral to the internet. In 1971, **Ray Tomlinson** sent the first email over ARPANET, introducing a new form of communication that would become ubiquitous in the digital age. In the 1980s, ARPANET transitioned to use the **TCP/IP protocol suite**, enabling the creation of a single, unified network of networks. This evolution allowed ARPANET to connect with other emerging networks, eventually forming the global internet.

The cultural impact of ARPANET—and later the internet—has been profound. It transformed the way people access information, communicate, and interact, breaking down geographical and social barriers. The network that began as a military experiment now underpins nearly every aspect of modern life, from commerce and education to entertainment and social connections.

ARPANET's success also highlighted the importance of **government-funded research** in driving technological innovation. By investing in foundational technologies, ARPA catalyzed a series of breakthroughs that would be commercialized and expanded upon by private industry. This partnership between public and private sectors has been a defining feature of the internet's development, ensuring its continued growth and adaptability.

Chapter Summary

The creation of ARPANET in 1969 marked the first step toward the modern internet, connecting distant computers through a decentralized, packet-switched network. This groundbreaking system revolutionized communication, enabling real-time collaboration and laying the foundation for a global network of interconnected devices. ARPANET's legacy extends far beyond its original purpose, shaping the way people access information, communicate, and innovate. It stands as a testament to the power of visionary thinking and collaborative research in driving transformative technological change.

Next Chapter

As the internet connected people and systems, another revolutionary technology was emerging that would change the way scientists studied life itself. In the next chapter, we'll explore the development of the **polymerase chain reaction (PCR)**, a technique that enabled scientists to amplify DNA and unlock new possibilities in genetic research, diagnostics, and biotechnology.

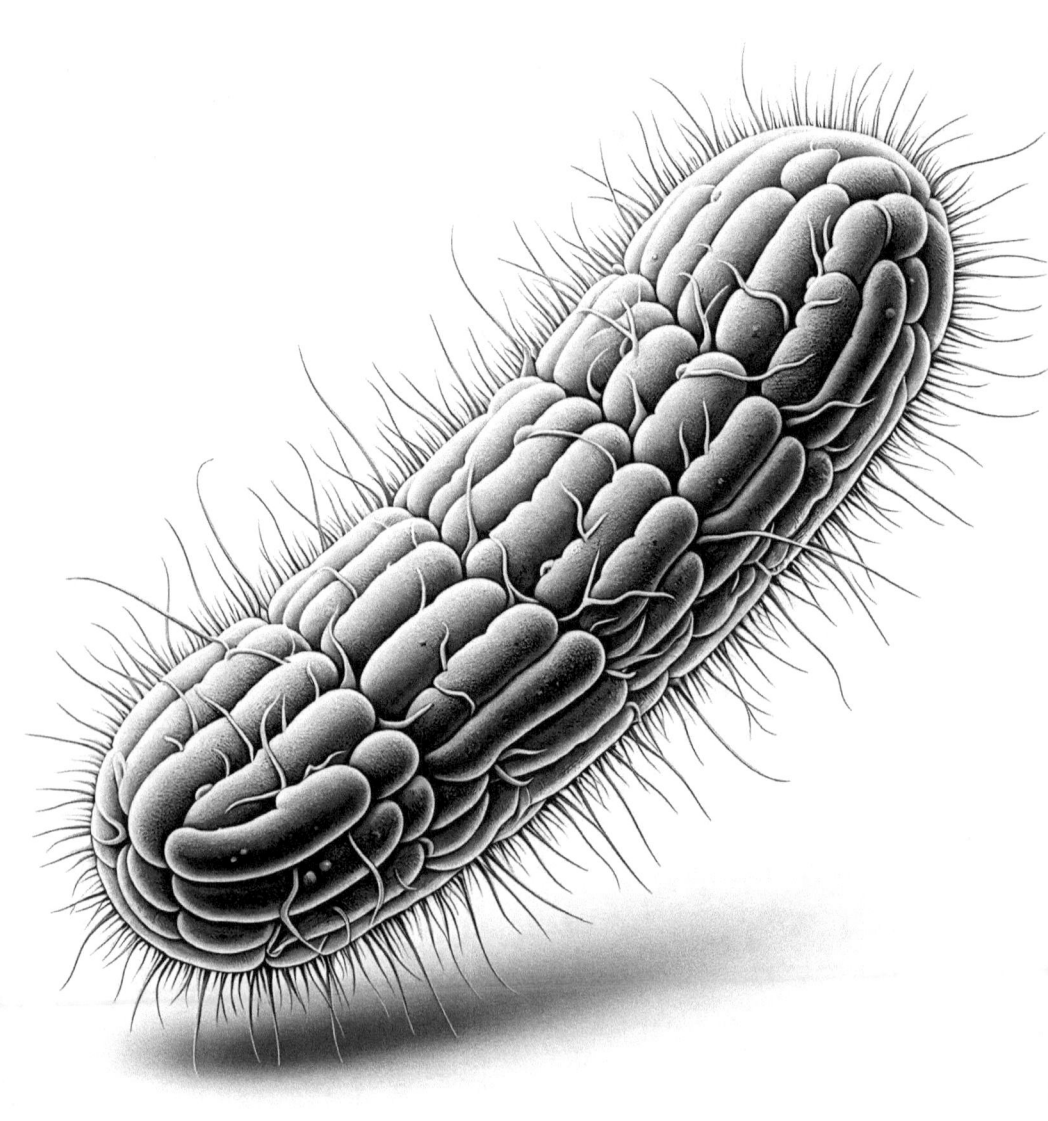

037 Polymerase Chain Reaction
A Scientist's Mistake Enables DNA Replication

The **polymerase chain reaction (PCR)**, one of the most transformative techniques in modern molecular biology, was developed in the early 1980s by **Kary Mullis**, a biochemist working at Cetus Corporation. PCR emerged from Mullis's quest to solve a problem: how to amplify tiny amounts of DNA to study them in detail. At the time, DNA analysis was labor-intensive and limited by the small amounts of genetic material that could be isolated from biological samples. Mullis's breakthrough came during a road trip in 1983, when he conceptualized a method to use a series of **temperature changes** to amplify DNA exponentially.

The core of PCR lies in the use of **Taq polymerase**, an enzyme that can withstand the high temperatures required to separate DNA strands during the reaction. Taq polymerase was discovered in the 1970s in the thermophilic bacterium **Thermus aquaticus**, which thrives in hot springs. However, Mullis initially overlooked its potential for PCR and used a less effective enzyme. It was only through iterative experimentation, and with input from colleagues, that the robustness of Taq polymerase became apparent. This "mistake" in the initial choice of enzymes underscores how trial and error can lead to innovation.

In 1985, Mullis and his team published the first demonstration of PCR, showing how DNA sequences could be amplified millions of times within hours. This method revolutionized molecular biology by providing a fast, efficient, and highly reproducible way to amplify genetic material, enabling breakthroughs in research, medicine, and forensic science.

Importance and Impact

PCR transformed molecular biology by making DNA amplification simple, rapid, and widely accessible. Before PCR, scientists needed large amounts of DNA for experiments, which often required painstaking extraction from samples. PCR allowed researchers to start with minuscule amounts of DNA and generate billions of copies, making previously impossible studies routine.

One of PCR's most immediate impacts was in **medical diagnostics**. It enabled the detection of genetic mutations associated with diseases like **cystic fibrosis** and **Huntington's disease**, facilitating earlier and more accurate diagnoses. PCR also became a cornerstone of **infectious disease testing, allowing scientists to detect pathogens like HIV, hepatitis, and, more recently, SARS-CoV-2 (the virus responsible for COVID-19). Its speed and sensitivity made PCR an invaluable tool in controlling outbreaks and guiding public health responses.

PCR also revolutionized **forensic science**, becoming a standard method for analyzing DNA evidence. Tiny amounts of biological material—such as a single drop of blood or a strand of hair—could now be amplified and analyzed, transforming how criminal cases were solved. The ability to identify individuals based on their DNA has had a profound impact on the criminal justice system, exonerating the innocent and securing convictions with unprecedented reliability.

In research, PCR has been essential for mapping genomes, studying gene expression, and engineering new genetic constructs. It has enabled advancements in **genomics**, **gene editing**, and **synthetic biology**, opening new frontiers in our understanding of life and the manipulation of biological systems. PCR is also used in **evolutionary biology** to analyze ancient DNA, offering insights into the genetics of extinct species like Neanderthals and woolly mammoths.

The invention of PCR exemplifies how a simple but powerful idea can change the course of science and society. Mullis's recognition of the method's potential, combined with the subsequent refinement of the technique by other scientists, underscores the collaborative nature of innovation. PCR's versatility and reliability have made it one of the most widely used tools in modern biology, with applications ranging from cancer research to agriculture.

Chapter Summary

The polymerase chain reaction, developed by Kary Mullis in 1983, revolutionized molecular biology by enabling the rapid amplification of DNA. This technique transformed medical diagnostics, forensic science, and genetic research, making DNA analysis accessible and routine. PCR's impact extends to diverse fields, including infectious disease control, evolutionary biology, and synthetic biology, demonstrating its profound influence on both science and society. The method's development highlights the importance of curiosity, iteration, and collaboration in driving scientific breakthroughs.

Next Chapter

As PCR unlocked the secrets of DNA, another innovation was making waves in imaging and diagnostics. In the next chapter, we'll explore the development of **quantum dots**, nanoscale particles with extraordinary optical properties that revolutionized imaging technologies and opened new possibilities in medical diagnostics and beyond.

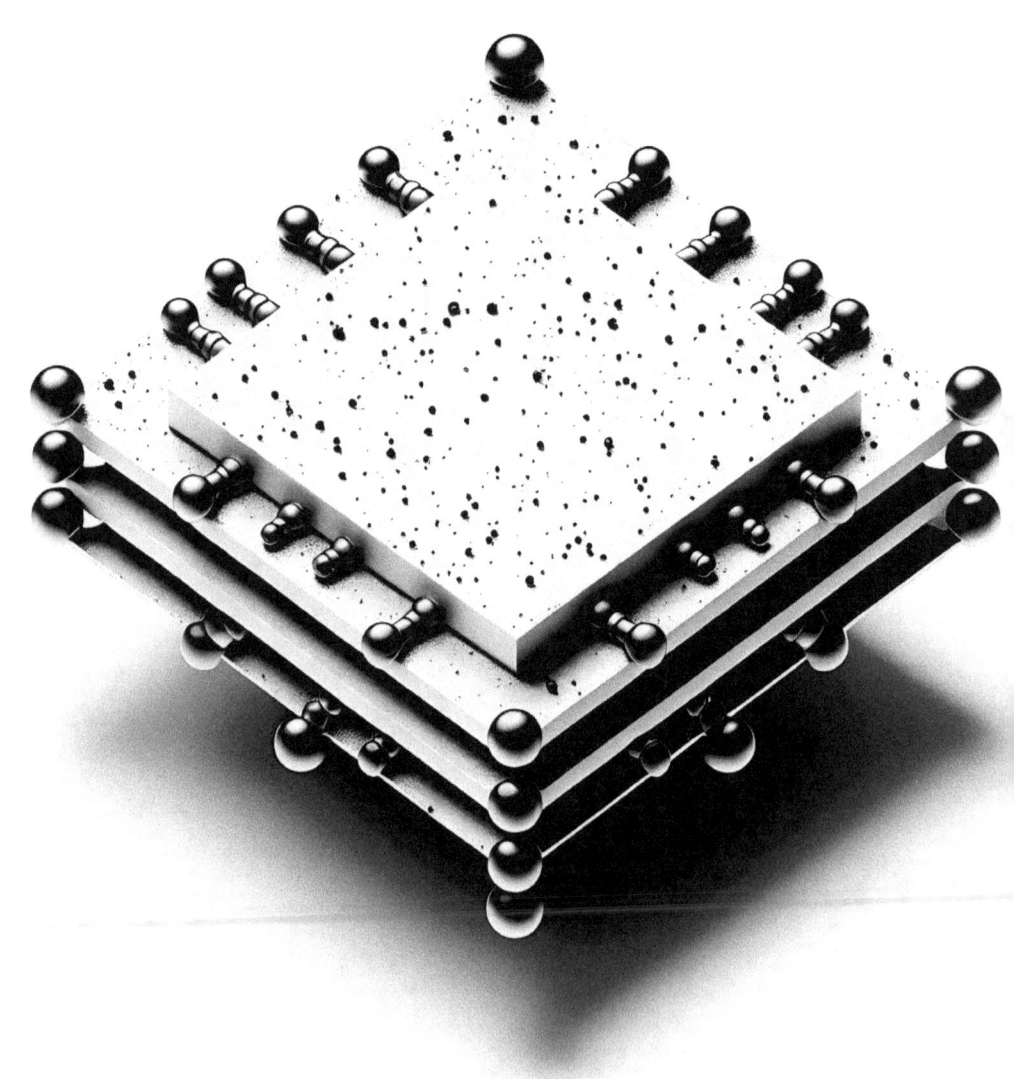

038 Quantum Dots
Nanoscale Color Revolution in Imaging

In the 1980s, a remarkable discovery in the field of nanotechnology unveiled a new type of material with extraordinary properties: **quantum dots**. These are **nanoscale semiconductor particles** that exhibit unique optical and electronic behaviors, arising from their quantum mechanical characteristics. The discovery of quantum dots was the result of decades of research into the behavior of materials at the atomic and molecular scale, where classical physics gives way to quantum effects.

The foundational work began with **Alexei Ekimov** in 1981, who demonstrated quantum confinement in a glass matrix. Soon after, **Louis Brus** created colloidal quantum dots in a solution, proving their existence and stability. These breakthroughs showed that quantum dots could absorb and emit light in highly specific wavelengths, depending on their size. By changing the particle's size, scientists could "tune" its color, creating a spectrum of vibrant hues that made quantum dots highly versatile for imaging and display technologies.

The practical potential of quantum dots became evident in the 1990s, when researchers began exploring their use in **biological imaging**. Their intense, size-dependent fluorescence was ideal for labeling cells, proteins, and DNA, allowing scientists to visualize molecular interactions with unprecedented clarity. By the early 2000s, quantum dots had become indispensable in both research and commercial applications, from medical diagnostics to cutting-edge display screens.

Importance and Impact

Quantum dots revolutionized imaging by offering unparalleled brightness, precision, and durability. In **biological research**, their fluorescent properties made them ideal for tracking cellular processes, identifying disease markers, and studying complex molecular systems. Traditional dyes and imaging agents often faded quickly or produced broad, overlapping signals, but quantum dots provided stable, sharp, and tunable emission spectra, allowing for more detailed and reliable observations.

One of the most transformative applications of quantum dots has been in **medical diagnostics**. Their ability to attach to specific biomolecules, such as antibodies or DNA sequences, enables highly targeted imaging of disease sites, including cancerous tumors. Quantum dot-enabled imaging has improved the accuracy of diagnoses, facilitated earlier detection of diseases, and enhanced the precision of treatments like **targeted drug delivery**.

Quantum dots also made a significant impact in **display technology**, where their vibrant and energy-efficient properties have been harnessed in modern televisions, monitors, and smartphone screens. **Quantum dot displays (QLEDs)** produce more vivid colors and consume less energy compared to traditional LED or OLED screens. This innovation has transformed the way people experience digital content, raising the standard for image quality in consumer electronics.

In addition to imaging and displays, quantum dots have opened new possibilities in **solar energy**. Their ability to absorb and convert sunlight into electricity with high efficiency makes them promising candidates for next-generation solar panels. This application could play a crucial role in addressing the global need for sustainable and renewable energy solutions.

The development of quantum dots also highlights the interdisciplinary nature of innovation. By combining advances in **physics**, **chemistry**, and **materials science**, researchers turned a theoretical concept into a practical technology with far-reaching implications. Their story illustrates how fundamental discoveries in quantum mechanics can lead to transformative tools that touch nearly every aspect of modern life.

Chapter Summary

Quantum dots, discovered in the 1980s, revolutionized imaging and diagnostics with their size-dependent optical properties. These nanoscale particles offered unprecedented precision in biological research, enabled breakthroughs in medical diagnostics, and transformed display technology with vibrant, energy-efficient screens. Their potential in renewable energy underscores their versatility and importance in addressing global challenges. The story of quantum dots is a testament to the power of interdisciplinary research and the transformative potential of nanotechnology.

Next Chapter

As quantum dots illuminated the microscopic world, another innovation was reshaping how sound was perceived by those who could not hear. In the next chapter, we'll explore the development of **cochlear implants**, devices that bypass the damaged parts of the ear to restore hearing and transform lives, exemplifying the profound intersection of technology and human health.

Cochlear Implants

Restoring Sound to the Deaf

The quest to restore hearing to those who are profoundly deaf has long challenged scientists and doctors. Early efforts in the 18th and 19th centuries focused on mechanical devices like ear trumpets and later, electrical stimulation of the auditory nerve. However, these methods provided limited results, offering only rudimentary sound awareness. By the mid-20th century, advances in electronics and neuroscience opened new possibilities for directly interfacing with the auditory system, leading to the concept of the **cochlear implant**.

The first rudimentary cochlear implants were developed in the 1950s and 1960s, with pioneers like **André Djourno**, **Charles Eyriès**, and **William House** experimenting with direct electrical stimulation of the auditory nerve. These early devices, while crude, demonstrated that electrical signals could produce the perception of sound in deaf individuals. By the 1970s, research intensified, leading to the development of multi-channel cochlear implants that could stimulate different parts of the cochlea, allowing for more nuanced sound perception. In 1978, **Graeme Clark**, an Australian surgeon, performed the first successful implantation of a multi-channel device, marking a significant milestone in the field.

Cochlear implants became commercially available in the 1980s, and their design continued to improve. By combining miniaturized electronics, advanced signal processing, and sophisticated surgical techniques, cochlear implants evolved into life-changing devices capable of providing a sense of hearing to those who had lost it entirely.

Importance and Impact

Cochlear implants transformed the lives of millions of individuals with profound hearing loss, offering them a way to perceive sound and engage with the world in ways that were previously impossible. Unlike hearing aids, which amplify sound, cochlear implants bypass damaged parts of the ear and directly stimulate the auditory nerve, enabling the brain to interpret sound signals. This technological breakthrough provided a solution for individuals who could not benefit from conventional hearing aids.

One of the most significant impacts of cochlear implants has been on **language development** in children born deaf. Early implantation allows children to develop speech and language skills at rates comparable to their hearing peers, profoundly improving their educational and social opportunities. For adults who lose their hearing later in life, cochlear implants restore access to sound, helping them regain independence and improve their quality of life.

Cochlear implants have also driven advancements in **neuroscience** and **neuroprosthetics**, providing insights into how the brain processes sensory information. The success of these devices has inspired the development of other neuroprosthetic technologies, such as retinal implants for vision restoration and brain-machine interfaces for individuals with paralysis. The cochlear implant remains one of the most successful examples of how technology can interface with the nervous system to restore lost functions.

Despite their transformative benefits, cochlear implants also sparked debates within the deaf community. Some critics argued that these devices promoted a medical model of deafness as a condition to be "fixed," potentially undermining Deaf culture and identity. These discussions highlighted the importance of balancing technological progress with cultural sensitivity and individual choice.

Today, cochlear implants continue to improve, with advancements in **wireless connectivity**, **sound processing algorithms**, and **bilateral implantation** enabling users to experience richer and more natural soundscapes. As the technology becomes more accessible, cochlear implants hold the promise of further breaking down barriers for individuals with hearing loss, ensuring their full participation in society.

Chapter Summary

Cochlear implants revolutionized the treatment of profound hearing loss by directly stimulating the auditory nerve, bypassing damaged parts of the ear. Pioneers like Graeme Clark and William House transformed experimental concepts into life-changing devices, enabling millions of people to perceive sound and engage with the world. These devices have profoundly impacted language development, quality of life, and the field of neuroprosthetics, while also sparking important cultural discussions. The cochlear implant exemplifies the transformative power of technology to restore human senses and bridge the gap between disability and possibility.

Next Chapter

As cochlear implants brought sound to the deaf, another innovation was restoring vision to the blind. In the next chapter, we'll explore the development of **bionic eyes**, which use advanced electronics and neuroprosthetics to provide partial vision to individuals with severe visual impairments, offering a glimpse of the future of sight restoration.

040 The bionic eye
Technology meets vision restoration

The quest to restore vision to those with severe visual impairments has fascinated scientists for centuries. While glasses and contact lenses have long corrected certain visual impairments, conditions like **retinitis pigmentosa** and **macular degeneration**, which damage the retina's light-sensitive cells, have left millions without effective solutions. The idea of creating a **bionic eye**, a device capable of bypassing damaged parts of the visual system to restore sight, emerged as advances in electronics and neuroscience converged in the late 20th century.

Early attempts to restore vision involved crude experiments with electrical stimulation. In the 1960s, researchers discovered that applying electrical currents to the optic nerve could produce rudimentary visual sensations, such as flashes of light. These experiments demonstrated that the brain could interpret electrical signals as visual information, paving the way for more sophisticated technologies. By the 1990s, researchers began developing the first **retinal implants**, devices designed to stimulate the remaining cells in the retina or bypass them entirely by transmitting signals directly to the optic nerve or visual cortex.

In 2013, the first commercially available bionic eye system, Argus II, was approved for use in the United States and Europe. Developed by Second Sight Medical Products, the **Argus II** system included a small camera mounted on a pair of glasses, which transmitted visual information to a chip implanted in the retina. Although the vision provided by the Argus II was rudimentary—limited to shapes, outlines, and contrasts—it represented a monumental step toward restoring functional vision to the blind.

Importance and Impact

The bionic eye marked a significant breakthrough in the field of **neuroprosthetics**, showcasing the potential of technology to restore lost senses. While the vision provided by these devices is not equivalent to natural sight, even limited visual restoration can profoundly improve quality of life. Recipients of bionic eye systems have reported being able to **navigate their surroundings**, recognize large objects, and, in some cases, read large text or distinguish faces.

The development of bionic eyes has also driven advancements in **neuroscience** and **brain-computer interfaces**, offering insights into how the brain processes visual information. These technologies are closely linked to ongoing research into the brain's adaptability, or **neuroplasticity**, as users learn to interpret the signals provided by their implants. The success of bionic eye systems has inspired researchers to develop similar technologies for other sensory and motor impairments, including hearing, touch, and movement.

Bionic eye systems also hold significant promise for future improvements. Researchers are working to enhance the resolution and field of view of these devices by integrating **higher-density electrode arrays** and **advanced imaging systems**. Efforts to develop cortical implants, which bypass the retina and stimulate the visual cortex directly, aim to provide vision to individuals whose optic nerves are damaged. The integration of **artificial intelligence (AI)** and machine learning into bionic eye systems may further improve image processing, enabling users to perceive more detailed and dynamic environments.

Despite these advances, challenges remain. Bionic eye systems are expensive, and access to these technologies is limited. Long-term studies are needed to understand their effectiveness and reliability fully. Additionally, ethical questions about the use of such technologies, particularly their implications for human enhancement, continue to spark debate. However, the progress made thus far underscores the transformative potential of merging biology and technology.

Chapter Summary

The bionic eye represents a groundbreaking achievement in vision restoration, offering new hope to individuals with severe visual impairments. By bypassing damaged parts of the visual system, devices like the Argus II have enabled recipients to perceive their surroundings and regain a degree of independence. This innovation has not only improved lives but also advanced the fields of neuroprosthetics and brain-computer interfaces. As technology continues to evolve, the bionic eye holds the promise of further breakthroughs in restoring vision and enhancing our understanding of the brain's capabilities.

Next Chapter

As bionic eyes restored vision, another innovation in communication was being forged under extraordinary circumstances. In the next chapter, we'll explore the story of the **Navajo Code Talkers** during World War II, whose ingenious use of their native language created an unbreakable code that revolutionized secure communication and influenced modern cryptographic methods.

041 The Unbreakable Code

Navajo Code Talkers and the Birth of Secure Communication

During **World War II**, secure communication became a critical challenge for the Allied forces. Encrypted messages were frequently intercepted by enemy forces, and despite advances in cryptographic techniques, codes could often be broken, leading to devastating consequences on the battlefield. In this context, the U.S. Marine Corps turned to an unconventional and ingenious solution: leveraging the **Navajo language** to create an unbreakable code.

The idea was the brainchild of **Philip Johnston**, a World War I veteran and the son of missionaries who had spent years living on the Navajo Nation. Johnston knew that Navajo, a complex and unwritten language, was spoken fluently by only a small number of people, making it an ideal foundation for a code. In 1942, he proposed his idea to military officials, who quickly realized its potential. The Marine Corps recruited **29 Navajo men**, later known as the **First 29 Code Talkers**, to develop a military code based on their language.

The Code Talkers devised a system that assigned Navajo words to represent military terms, such as using the word for "turtle" to signify a tank. They also created a phonetic alphabet, translating English letters into Navajo words. The resulting code was not only **fast** and **efficient** but also virtually indecipherable to anyone without an intimate knowledge of both Navajo and the code's structure. During their first deployment in the Pacific Theater, the Code Talkers proved their value, relaying critical messages under fire with unparalleled accuracy and speed.

Importance and Impact

The Navajo Code Talkers played a pivotal role in the Allied victories in the Pacific Theater, including battles such as **Iwo Jima**, **Guadalcanal**, and **Saipan**. Their contributions allowed military units to coordinate complex operations without fear of enemy interception, often turning the tide of battle. Unlike traditional encryption methods, which could be time-consuming and required machines, the Navajo code relied on human ingenuity, enabling real-time communication under the most demanding conditions.

The code's success lay in its **dual-layered security**. First, it used the Navajo language, which was incomprehensible to most non-Navajo speakers. Second, the additional coding layer—using Navajo words to represent specific military terms—rendered it even more difficult to decipher. Despite repeated attempts, the Japanese military never broke the Navajo code, making it one of the only truly unbreakable codes in modern military history.

Beyond their wartime contributions, the Code Talkers left a lasting legacy in the fields of **cryptography** and **secure communication**. Their innovative approach inspired subsequent developments in code design, including the use of culturally unique or context-specific encryption systems. The concept of combining linguistic complexity with layered encryption remains relevant in modern cybersecurity, where multifactor security systems are standard.

The story of the Navajo Code Talkers also highlights the broader contributions of Indigenous peoples to science, technology, and innovation. Despite facing systemic discrimination and exclusion from mainstream society, these men brought their unique skills and cultural heritage to bear on a problem of global significance. Their achievements were not officially recognized until decades later, but their story has become a powerful symbol of resilience and ingenuity.

Today, the legacy of the Code Talkers continues to inspire advancements in secure communication technologies, including **quantum cryptography** and **artificial intelligence-based encryption**. Their story serves as a reminder of the potential of human creativity and collaboration to solve complex problems, even under the most challenging circumstances.

Chapter Summary

The Navajo Code Talkers of World War II created an unbreakable communication system that played a crucial role in the Allied victory in the Pacific. By combining the complexity of the Navajo language with innovative coding techniques, they ensured secure communication under fire, inspiring modern approaches to encryption. Their contributions underscore the importance of diversity and ingenuity in solving critical challenges and highlight the enduring legacy of their work in secure communication.

Next Chapter

As the Code Talkers used language to solve communication challenges, another accidental discovery in chemistry was reshaping science. In the next chapter, we'll explore the **discovery of fullerenes**, a new form of carbon with extraordinary properties that revolutionized nanotechnology and materials science.

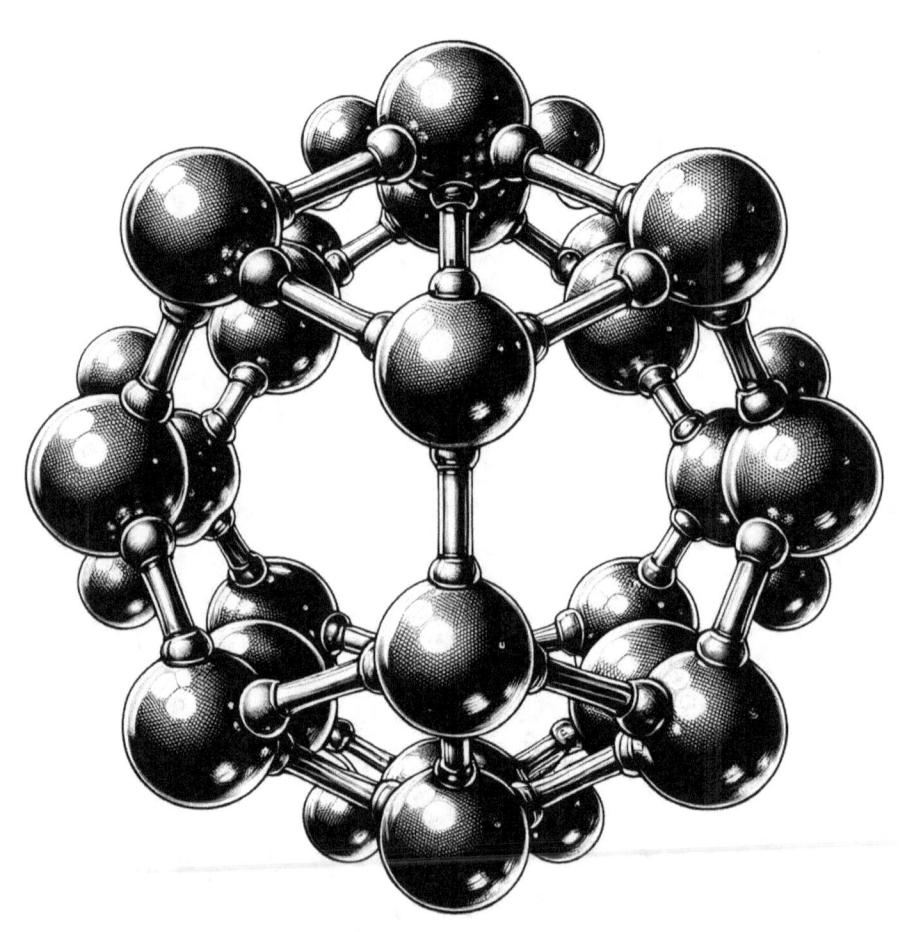

042 The Accidental Discovery of Fullerenes
Carbon's Unexpected Shape

In 1985, a serendipitous discovery during a seemingly routine experiment would transform the field of **nanotechnology** and materials science. Researchers **Harold Kroto**, **Richard Smalley**, and **Robert Curl** were studying carbon molecules using laser vaporization techniques to simulate the conditions in stars. Their goal was to understand how long carbon chains formed in space. However, what they discovered instead was a completely new molecular structure: a spherical, cage-like form of carbon comprising 60 atoms. They named it **buckminsterfullerene**, or **buckyball**, after architect **Buckminster Fuller**, whose geodesic domes resembled the molecule's shape.

The buckyball was the first member of a new class of carbon molecules known as **fullerenes**, characterized by their unique arrangement of carbon atoms into hexagons and pentagons. This discovery was groundbreaking because it revealed a previously unknown form of pure carbon, distinct from graphite and diamond. Fullerenes exhibited remarkable stability and unusual properties, opening up new avenues for research and application. The team's discovery earned them the **1996 Nobel Prize in Chemistry**, and fullerenes became a foundational element in the burgeoning field of nanotechnology.

Importance and Impact

The discovery of fullerenes marked a turning point in materials science, revealing the versatility and potential of carbon at the nanoscale. One of their most remarkable properties was their **strength and resilience**. Despite being incredibly lightweight, fullerenes could withstand enormous pressure, making them ideal candidates for applications in structural materials, coatings, and nanocomposites.

In **medicine**, fullerenes opened new possibilities for drug delivery and diagnostics. Their hollow, cage-like structure allowed researchers to trap and transport molecules, such as drugs or imaging agents, to targeted locations in the body. This potential has been explored for treating diseases like cancer and for delivering antioxidants to counteract oxidative stress.

Fullerenes also contributed to advancements in **electronics** and **energy storage**. Their ability to conduct electricity and resist degradation under harsh conditions made them valuable for developing more efficient solar cells, batteries, and superconductors. Research into fullerene-based materials continues to push the boundaries of renewable energy technologies, offering solutions for reducing reliance on fossil fuels.

The discovery of fullerenes also inspired the exploration of other carbon-based nanostructures, such as **carbon nanotubes** and **graphene**, which have become central to nanotechnology. Carbon nanotubes, for instance, are incredibly strong and conductive, finding applications in everything from aerospace engineering to flexible electronics. Graphene, a single layer of carbon atoms arranged in a hexagonal lattice, has revolutionized materials science with its extraordinary strength, transparency, and electrical properties.

Beyond their scientific and technological impact, fullerenes have deepened our understanding of **chemical bonding and molecular architecture**, demonstrating how carbon's unique versatility underpins much of modern science. The story of their discovery also underscores the value of curiosity-driven research. Kroto, Smalley, and Curl's original goal had nothing to do with nanotechnology or materials science, but their openness to unexpected results led to a discovery that transformed multiple fields.

Chapter Summary

The accidental discovery of fullerenes in 1985 revealed a new form of carbon with extraordinary properties, from strength and stability to electrical conductivity. Fullerenes revolutionized nanotechnology, enabling breakthroughs in medicine, energy storage, and materials science. This discovery also laid the foundation for the exploration of other carbon nanostructures, such as nanotubes and graphene, which continue to shape the future of technology. The story of fullerenes highlights the transformative power of curiosity-driven research and the potential of unexpected discoveries to drive innovation.

Next Chapter

As fullerenes reshaped our understanding of molecular structures, another scientific curiosity involving amphibians would unexpectedly advance our knowledge of neural connectivity. In the next chapter, we'll explore the strange story of **"telepathic frogs"**, where an unusual experiment led to breakthroughs in neuroscience and the early foundations of brain-machine interfaces.

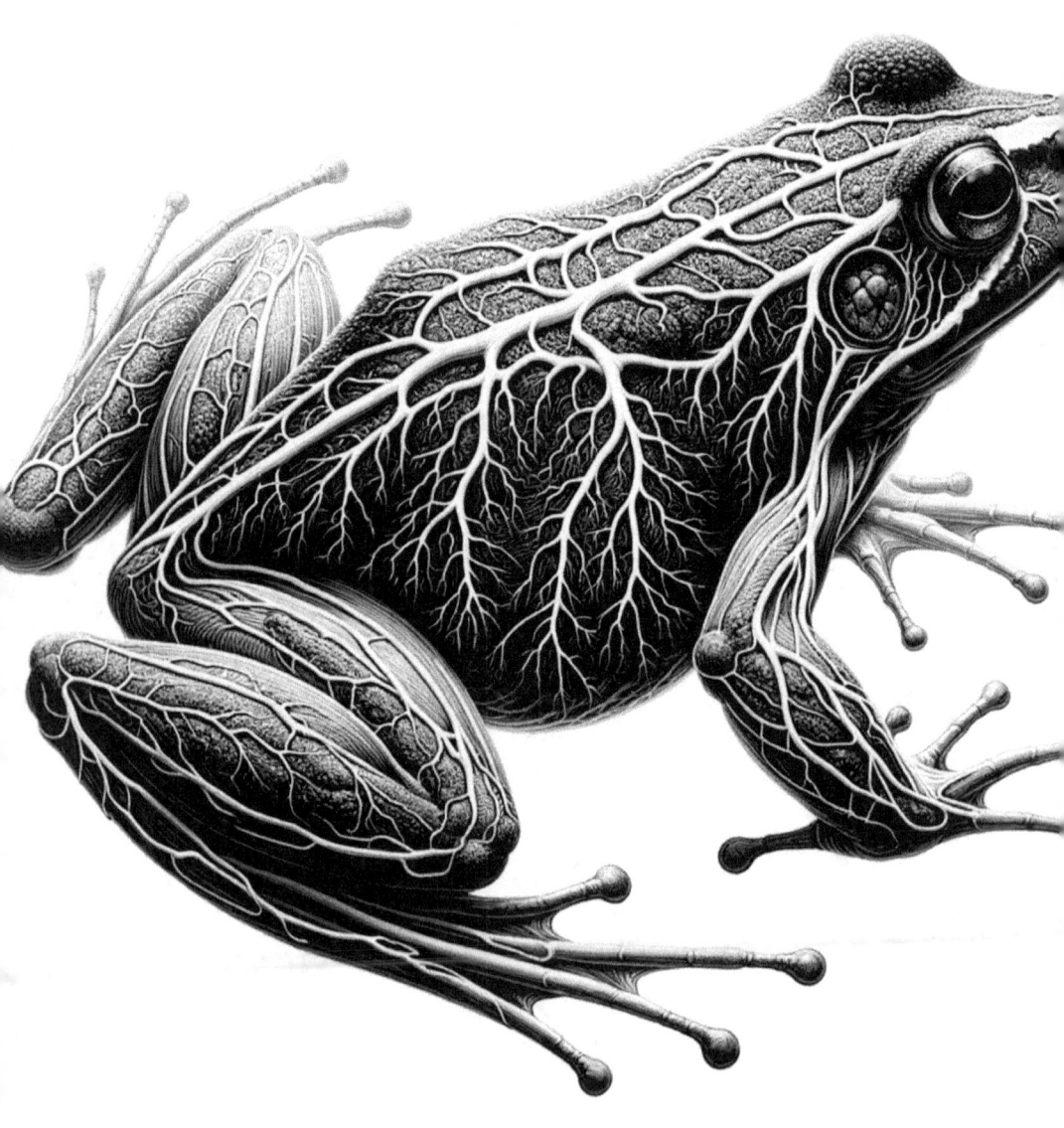

043 The Telepathic Frogs
A Misunderstood Experiment That Advanced Neuroscience

In the early 20th century, neuroscience was still in its infancy. Scientists were grappling with fundamental questions about how the brain and nervous system worked, and experiments often relied on trial and error. Amid this exploratory era, a peculiar experiment in the 1930s involving **frog nerves** unexpectedly contributed to our understanding of **neural connectivity and signal transmission**, laying the groundwork for **brain-machine interfaces** in later decades.

The experiment was conducted by **Otto Loewi**, a Nobel Prize-winning physiologist renowned for discovering chemical transmission between neurons. Loewi aimed to demonstrate how electrical signals could propagate through neural circuits in amphibians. To do this, he wired two frog leg muscles together using electrodes, stimulating one leg while observing the other's response. To his surprise, the second leg appeared to move in tandem with the first, as though the frogs were communicating telepathically.

The results were misinterpreted at the time, with some observers describing the phenomenon as a form of "nervous telepathy." However, further investigation revealed a more grounded explanation: the electrodes were transmitting **electrical signals** between the two frogs, mimicking the natural flow of signals in the nervous system. This realization opened the door to deeper explorations of how artificial stimulation could interact with biological systems, an insight that would prove pivotal in neuroscience and robotics.

Importance and Impact

The "telepathic frogs" experiment, while misunderstood at first, became an important step in advancing our understanding of the nervous system's electrical properties. It confirmed that **neural circuits** could be manipulated and studied using external electrical inputs, inspiring subsequent research into the interface between technology and biology. These early insights formed the basis for what would later become **neuroprosthetics** and **brain-machine interfaces (BMIs)**.

One of the most significant applications of this research was the development of devices that could bypass damaged nerves or replicate their functions. Technologies like **cochlear implants** and **deep**

brain stimulation (used to treat conditions like Parkinson's disease) are direct descendants of experiments like Loewi's. By showing that electrical signals could interact with neural pathways, the study paved the way for innovative solutions to neurological disorders and injuries.

The principles uncovered in the "telepathic frogs" experiment also influenced the field of **robotics**. Early cyberneticists drew on these findings to develop robotic systems that could mimic biological processes, including the use of sensors and feedback loops to control movement. This line of research eventually led to the creation of **brain-controlled prosthetics**, allowing individuals with amputations to control artificial limbs using their thoughts.

Beyond its immediate scientific applications, the experiment sparked broader philosophical questions about the nature of communication and the interplay between biology and technology. It highlighted the potential for artificial systems to replicate or augment natural processes, an idea that continues to drive innovation in fields like **bioengineering** and **artificial intelligence**.

Despite its initial misinterpretation, the "telepathic frogs" experiment serves as a testament to the importance of open-mindedness in scientific inquiry. What began as an attempt to study simple neural circuits evolved into a cornerstone of modern neuroscience, illustrating how even misunderstood results can lead to transformative discoveries.

Chapter Summary

The "telepathic frogs" experiment of the 1930s, initially misinterpreted as a form of nervous telepathy, revealed fundamental truths about how neural circuits respond to artificial stimulation. This groundbreaking work inspired the development of neuroprosthetics, brain-machine interfaces, and robotic systems that mimic biological functions. The experiment's legacy underscores the importance of curiosity-driven research and the potential of unexpected findings to revolutionize science and technology.

Next Chapter

As experiments with frog nerves illuminated the principles of neural connectivity, another breakthrough in nanotechnology was quietly taking shape. In the next chapter, we'll explore the development of **electron-beam lithography**, a technique that enabled the creation of nanometer-scale devices and paved the way for the microelectronics revolution.

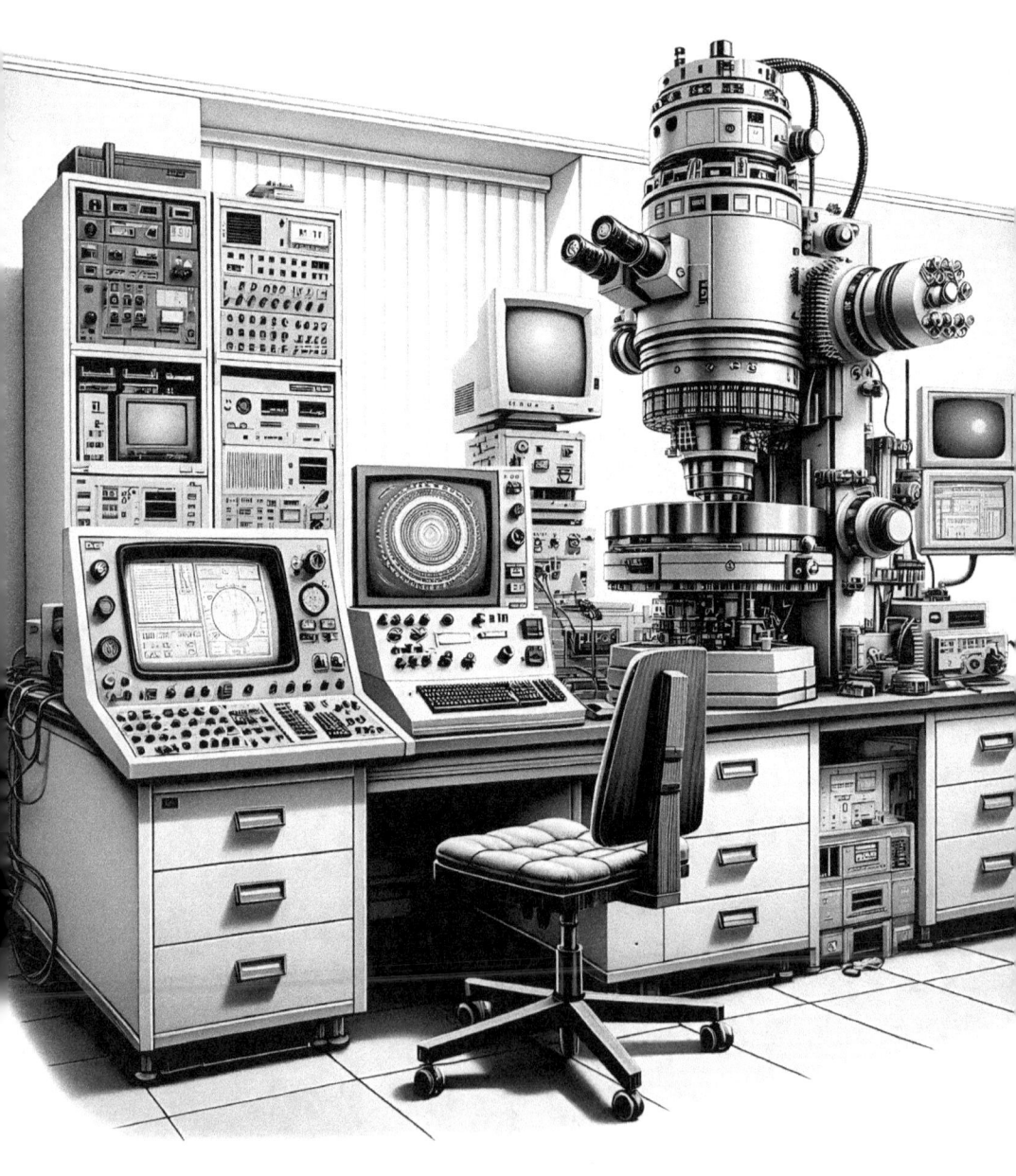

044 The Shrinking Machine
Electron-Beam Lithography and the Path to Nanodevices

As electronics advanced in the mid-20th century, the need for ever-smaller and more efficient components drove innovation in **fabrication techniques**. The invention of the **integrated circuit (IC)** in the late 1950s had revolutionized computing, but by the 1960s, scientists and engineers faced a new challenge: how to manufacture devices at even smaller scales. Traditional photolithography, which used light to etch patterns onto silicon wafers, was reaching its resolution limits due to the physical constraints of light's wavelength.

The solution came in the form of **electron-beam lithography (EBL)**, a technique that used a focused beam of electrons instead of light to create nanoscale patterns. First developed in the late 1960s by researchers at institutions like **Bell Labs** and **IBM**, EBL allowed for the creation of features as small as a few nanometers. The technique leveraged the shorter wavelength of electrons compared to visible light, enabling unprecedented precision in the etching of circuits and structures.

Although initially slow and expensive, electron-beam lithography was a game-changer for **nanotechnology** and **microelectronics**. It became a critical tool for prototyping advanced semiconductor devices, pushing the boundaries of miniaturization and paving the way for technologies like the microprocessor, high-density memory, and nanosensors.

Importance and Impact

The development of electron-beam lithography was a key milestone in the race to miniaturize electronic components. By enabling the fabrication of features far smaller than those achievable with photolithography, EBL unlocked new levels of **performance** and **efficiency** in semiconductor devices. This advancement directly supported the exponential growth described by **Moore's Law**, which predicted the doubling of transistor density on chips approximately every two years.

EBL's precision also made it indispensable for research in **nanotechnology**. Scientists used the technique to fabricate nanoscale devices and structures, including quantum dots, nanowires, and nanosensors. These breakthroughs laid the groundwork for modern applications in fields like **medicine, energy**, and **communications**, from drug delivery systems to high-efficiency solar cells.

In addition to its role in electronics, EBL enabled advancements in **fundamental physics** and materials science. Researchers could use the technique to create custom experimental setups at the atomic scale, investigating phenomena like superconductivity, quantum tunneling, and plasmonics. These studies deepened our understanding of the natural world and informed the design of novel materials with tailored properties.

Electron-beam lithography also played a critical role in the development of **extreme ultraviolet lithography (EUVL)**, a next-generation fabrication technique now used for mass-producing chips with nanometer-scale features. By serving as a prototyping tool, EBL helped refine the processes and designs necessary for scaling up nanofabrication, ensuring its continued relevance in the semiconductor industry.

While EBL remains a specialized tool due to its high cost and slow throughput, its impact on **research** and **innovation** cannot be overstated. From enabling the production of cutting-edge electronics to facilitating breakthroughs in nanotechnology, EBL exemplifies how precision tools can drive progress across multiple scientific disciplines.

Chapter Summary

Electron-beam lithography revolutionized fabrication by enabling the creation of nanoscale devices and circuits, surpassing the resolution limits of traditional photolithography. This innovation played a critical role in advancing microelectronics, supporting the exponential growth of semiconductor technology, and paving the way for breakthroughs in nanotechnology. EBL's legacy lies in its precision and versatility, demonstrating how advancements in fabrication can catalyze progress in fields ranging from physics to medicine.

Next Chapter

As electron-beam lithography refined the art of fabrication, engineers were also harnessing the power of vibrations in surprising ways. In the next chapter, we'll explore the use of **artificial earthquakes** in designing safer buildings and how these man-made tremors reshaped urban planning in earthquake-prone areas.

045 Artificial Earthquakes
The Hidden Role of Vibrations in Building Modern Cities

Earthquakes have long been one of humanity's most feared natural disasters, capable of toppling cities and leaving devastating consequences in their wake. As urban centers expanded in the 20th century, the need to build **earthquake-resistant structures** became increasingly urgent. While seismologists made strides in understanding how natural earthquakes occurred, engineers began to explore how **man-made vibrations**—or **artificial earthquakes**—could simulate seismic activity to test and improve building designs.

The concept of generating artificial earthquakes dates back to the mid-20th century, when researchers sought to recreate seismic conditions in controlled environments. Early efforts involved mechanical shakers that applied vibrations to small models of buildings, helping engineers observe how different structures responded to stress. By the 1960s, advances in hydraulic and electronic systems enabled the creation of **shake tables**, platforms capable of mimicking the motions of actual earthquakes.

In 1972, Japan introduced the world's first **large-scale shake table**, capable of testing full-sized buildings. Known as the **E-Defense Facility**, it allowed engineers to simulate earthquakes with remarkable accuracy, using real-world seismic data to replicate the shaking patterns of historic quakes. These artificial tremors provided critical insights into how buildings behaved during seismic events, leading to innovations in structural design and urban planning.

Importance and Impact

The use of artificial earthquakes revolutionized the field of **seismic engineering** by enabling scientists and architects to test building designs under realistic conditions. This approach provided a wealth of data on how structures responded to various types of ground motion, from short, sharp jolts to long, rolling tremors. These insights led to the development of **earthquake-resistant technologies**, including **base isolators**, which allow buildings to move independently of the ground, and **dampers**, which absorb and dissipate seismic energy.

Artificial earthquakes also played a key role in updating **building codes** worldwide. Engineers used data from shake table tests to establish minimum safety standards for structures in earthquake-prone regions. These codes have significantly reduced casualties and property damage during seismic events, making cities safer for millions of people.

Beyond buildings, artificial earthquakes have been used to study the behavior of **infrastructure**, such as bridges, dams, and pipelines, under seismic stress. This research has led to innovations like flexible joints for pipelines, ensuring the continued flow of water and gas during emergencies. Shake table tests have also informed the design of **nuclear power plants**, ensuring their resilience to extreme seismic events.

Artificial seismic testing has even extended to **historical preservation**, helping engineers devise ways to protect ancient structures from earthquakes. In places like Italy and Greece, researchers have used shake tables to study how heritage sites might fare in future quakes, developing restoration techniques that preserve these cultural landmarks.

The success of artificial earthquakes highlights the interplay between science and engineering in solving real-world problems. By combining seismic data with cutting-edge technology, researchers have created tools that not only mitigate the effects of natural disasters but also inspire innovative approaches to urban planning and construction.

Chapter Summary

Artificial earthquakes transformed the study of seismic activity, enabling engineers to test and refine building designs under realistic conditions. Shake table experiments have driven the development of earthquake-resistant technologies, updated building codes, and ensured the safety of critical infrastructure. By simulating the destructive power of nature, artificial earthquakes have made cities more resilient and inspired a new era of innovation in urban planning and construction.

Next Chapter

As engineers used vibrations to protect cities, another revolutionary idea was being tested in robotics. In the next chapter, we'll explore the creation of **cybernetic turtles**, the world's first robotic animals, and how these early experiments in machine intelligence laid the groundwork for modern robotics and artificial intelligence.

046 The Cybernetic Turtle
The First Robot Animals and the Origins of Robotics

In the aftermath of World War II, the field of **cybernetics**—the study of control systems in living organisms and machines—emerged as a groundbreaking interdisciplinary science. Cybernetics sought to understand how feedback loops, sensors, and actuators could replicate biological behaviors in mechanical systems. One of the earliest and most intriguing experiments in this field came in the form of **robotic animals**: mechanical creatures designed to mimic the behavior of real animals. Among these, the **cybernetic turtles** created in the late 1940s by British neuroscientist **William Grey Walter** became iconic.

Walter's goal was to explore the relationship between simple neural networks and complex behaviors. He built a series of small robots, which he called **tortoises** due to their slow, deliberate movements. These robots were equipped with rudimentary sensors, motors, and a feedback system that allowed them to react to light and obstacles in their environment. Despite their simplicity, the cybernetic turtles displayed surprisingly lifelike behaviors, such as seeking light sources and avoiding collisions.

The turtles' behavior fascinated scientists and the public alike, as they appeared to exhibit goal-directed actions and even decision-making. Walter's creations demonstrated that complex behavior could arise from simple rules, laying the groundwork for future advancements in **robotics, artificial intelligence**, and **neuroscience**.

Importance and Impact

The cybernetic turtles were a profound demonstration of how machines could mimic biological processes. By showing that a small network of electrical components could replicate behaviors like **navigation** and **obstacle avoidance**, Walter's work bridged the gap between biology and engineering. This insight inspired subsequent generations of researchers to explore how artificial systems could simulate—and even surpass—biological intelligence.

One of the turtles' most important contributions was in advancing the concept of **autonomy in robotics**. Prior to Walter's work, machines were generally thought of as tools that required human control. The turtles, however, operated independently, responding to their environment in real-time without external guidance. This autonomy became a cornerstone of modern robotics, enabling the development of **self-driving cars**, **drones**, and **robotic explorers**.

The principles demonstrated by the cybernetic turtles also influenced the design of **neural networks**, the computational models that power much of today's artificial intelligence. Walter's experiments showed that simple, interconnected systems could produce emergent behaviors, a concept that underpins machine learning and AI research. His work anticipated ideas that would later revolutionize fields like **robotic swarm intelligence**, where groups of simple robots collaborate to perform complex tasks.

In addition to their scientific impact, the turtles captured the public's imagination, sparking widespread interest in robotics and AI. Walter showcased his creations at exhibitions and public events, inspiring curiosity about the potential of machines to imitate life. This cultural fascination with robotic animals paved the way for projects like **robotic pets** and **bio-inspired robots**, which continue to explore the boundary between the artificial and the organic.

Today, the legacy of the cybernetic turtles is evident in the ongoing development of biomimetic robots, which replicate the movements and behaviors of animals to navigate complex environments. From snake-like robots used in search-and-rescue missions to robotic fish that monitor marine ecosystems, these innovations owe their conceptual roots to Walter's pioneering work.

Chapter Summary

William Grey Walter's cybernetic turtles demonstrated how simple mechanical systems could mimic complex biological behaviors, laying the foundation for modern robotics and artificial intelligence. By pioneering the concept of autonomy in machines, Walter's work inspired advancements in neural networks, swarm intelligence, and biomimetic robotics. The turtles' legacy extends far beyond their simple design, shaping the trajectory of robotics and sparking enduring fascination with the intersection of biology and technology.

Next Chapter

As cybernetic turtles explored autonomy in machines, another revolutionary idea was taking shape in renewable energy. In the next chapter, we'll examine the cutting-edge concept of **harnessing pho-**

tosynthesis to generate electricity, a breakthrough that promises to merge biology and technology for sustainable energy solutions.

047 Plants That Power
Harnessing Photosynthesis for Renewable Energy

For centuries, humanity has relied on plants as sources of food, medicine, and materials. However, in recent decades, researchers began exploring a more futuristic role for plants: harnessing their ability to perform **photosynthesis** to generate **renewable energy**. Photosynthesis, the process by which plants convert sunlight into chemical energy, is one of nature's most efficient energy systems. Scientists realized that if this process could be adapted to produce electricity, it could offer a sustainable and environmentally friendly alternative to fossil fuels.

The concept of using photosynthesis for energy production gained traction in the late 20th century, as concerns about climate change and energy scarcity prompted the search for cleaner technologies. Early experiments involved integrating **chlorophyll**, the pigment responsible for capturing sunlight, into artificial systems to mimic photosynthesis. While promising, these attempts faced significant challenges in efficiency and scalability.

A breakthrough came in the early 21st century, when researchers began developing **living bio-photovoltaic systems**, which used actual plants or algae to generate electricity. By connecting plants to electrodes, scientists discovered they could capture the electrons produced during photosynthesis. This approach transformed plants into living solar panels, capable of producing energy while continuing their natural biological functions.

Importance and Impact

Harnessing photosynthesis for renewable energy has the potential to revolutionize how we think about energy generation. Unlike traditional solar panels, which require rare materials and complex manufacturing processes, bio-photovoltaic systems use readily available, biodegradable components. This makes them a more **sustainable and cost-effective solution**, particularly for communities with limited access to traditional energy infrastructure.

One of the most promising applications of this technology is in **micro-energy systems**, where small amounts of electricity generated by plants can power low-energy devices like sensors or LED lights. For instance, researchers have developed systems where houseplants power environmental sensors, creating a self-sustaining loop that monitors air quality or soil moisture without the need for batteries or external power sources.

Beyond small-scale applications, bio-photovoltaics could play a critical role in the future of **smart agriculture**. By integrating energy-generating plants into farming systems, it's possible to create farms that are both food and energy producers. This dual-purpose approach could help address the growing demand for both resources in a rapidly expanding global population.

Harnessing photosynthesis for energy also has significant implications for **climate change mitigation**. Unlike conventional solar panels, bio-photovoltaic systems actively absorb carbon dioxide during photosynthesis, contributing to carbon sequestration. This dual function makes them a uniquely beneficial technology in the fight against global warming.

Although still in its early stages, the field of bio-photovoltaics has inspired a broader exploration of **nature-inspired energy technologies**, including artificial photosynthesis systems. By mimicking the efficiency of plants, scientists are developing synthetic systems capable of splitting water to produce hydrogen fuel, a potential game-changer in renewable energy storage.

The story of harnessing photosynthesis for energy highlights the potential of merging biology and technology. By leveraging nature's time-tested processes, researchers are paving the way for innovative solutions to some of the world's most pressing challenges, from energy scarcity to environmental degradation.

Chapter Summary

Harnessing photosynthesis for renewable energy represents a groundbreaking fusion of biology and technology. By capturing electrons generated during photosynthesis, researchers have developed bio-photovoltaic systems that turn plants into living solar panels. This technology offers sustainable solutions for energy generation, smart agriculture, and climate change mitigation, demonstrating the transformative potential of nature-inspired innovation. The ongoing exploration of bio-photovoltaics underscores the importance of interdisciplinary research in creating a cleaner, greener future.

Next Chapter

As plants began powering small-scale devices, scientists discovered that **bacteria** could also generate electricity in surprising ways. In the next chapter, we'll explore the development of **microbial fuel cells**, a technology that harnesses the natural metabolic processes of bacteria to create energy, promising a new frontier in renewable power generation.

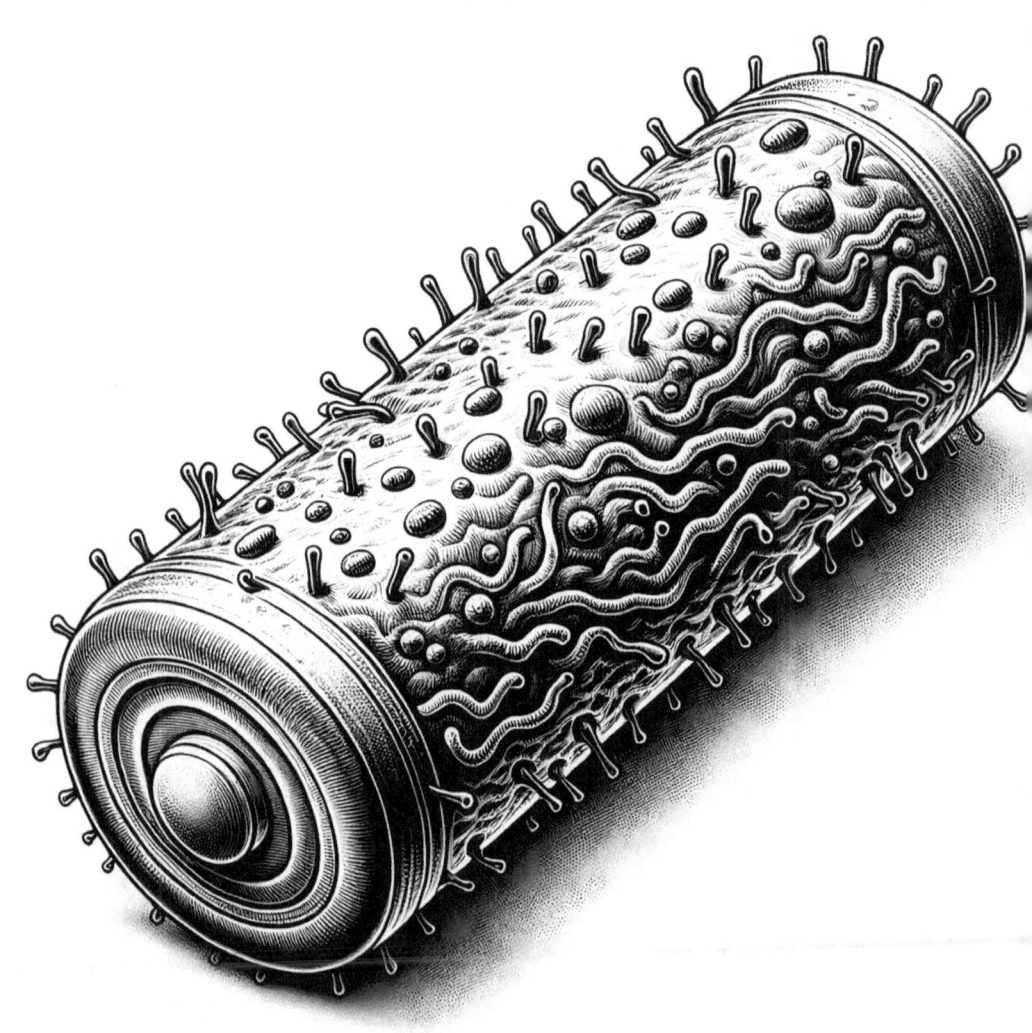

048 The Bacterial Battery
How Microbes Accidentally Became Electricians

In the 1980s, scientists examining microbial ecosystems in muddy riverbeds stumbled upon a bizarre phenomenon: certain bacteria were transferring electrons to metallic surfaces in their environment. This unexpected behavior, later termed **extracellular electron transfer**, appeared to be a strange quirk of microbial life. What made this discovery extraordinary was that it hinted at the ability of bacteria to generate electricity naturally, using their metabolic processes.

These electrogenic bacteria, including **Geobacter sulfurreducens** and **Shewanella oneidensis**, were first discovered by researchers studying how microbes survived in low-oxygen environments. Unlike humans, who use oxygen to "catch" electrons during respiration, these microbes relied on metals like iron and manganese. When electrodes were placed in the mud to study microbial behavior, the bacteria treated the electrodes as their new electron acceptors, creating measurable electric currents.

The first **microbial fuel cells (MFCs)** were little more than lab experiments. Mud was packed into small containers fitted with electrodes, and bacteria were allowed to metabolize the organic material in the mud. These setups produced small but steady amounts of electricity, demonstrating the potential for harnessing microbial metabolism as a power source. Though inefficient, the quirky notion of a "bacterial battery" captured the imagination of researchers.

This phenomenon became even more fascinating when scientists discovered that some bacteria could produce nanowires—microscopic filaments that conducted electricity. These **nanowires** allowed bacteria to transfer electrons over distances, connecting to electrodes or even forming networks with other bacteria. This behavior, resembling a natural electrical grid, challenged long-held assumptions about the capabilities of simple microorganisms.

As the field matured, researchers began exploring ways to optimize MFCs for practical use. Efforts focused on increasing efficiency, scaling up the systems, and finding real-world applications. What began as an oddity of nature had grown into a promising frontier for renewable energy research.

Importance and Impact

The discovery of electrogenic bacteria revolutionized how scientists thought about energy production and sustainability. Microbial fuel cells offered a unique dual benefit: they could generate electricity while simultaneously treating organic waste, making them a practical solution for wastewater treatment plants. By allowing bacteria to metabolize sewage and agricultural runoff, MFCs could produce power while cleaning polluted water.

Marine applications were among the first real-world uses of microbial fuel cells. In underwater environments, where traditional power sources are impractical, MFCs harnessed sediment-dwelling bacteria to power sensors. These systems provided a low-maintenance and sustainable energy source for monitoring ocean conditions, enabling long-term studies of marine ecosystems.

Electrogenic bacteria have also inspired innovations in **bioremediation**, where microbes are used to break down environmental pollutants like oil spills or heavy metals. The integration of microbial fuel cells into these processes allows cleanup efforts to generate electricity, making them more efficient and cost-effective.

Beyond their practical applications, microbial fuel cells have pushed the boundaries of **bioelectronics**. Researchers are working to engineer bacteria with enhanced electrical output, while others are combining microbial systems with solar panels to create hybrid energy devices. These innovations highlight the potential for integrating biological and traditional renewable energy sources.

Despite their promise, microbial fuel cells face challenges. Their energy output remains modest compared to other renewable technologies, and scaling them up for industrial use requires overcoming technical and economic hurdles. However, their environmental benefits and versatility make them an important component of the renewable energy landscape.

Chapter Summary

The discovery of electrogenic bacteria and their ability to generate electricity sparked the development of microbial fuel cells, which combine waste treatment with renewable energy production. From powering underwater sensors to enabling bioremediation, the bacterial battery showcases the transformative potential of nature-inspired innovation.

Next Chapter

As bacteria proved to be surprising sources of power, another unconventional approach to problem-solving emerged in computing. In the next chapter, we'll dive into the world of **liquid computers**, where swirling fluids and flowing dyes reveal unexpected ways to tackle complex challenges.

049 Liquid Computers
Solving Problems with Sloshing Solutions

In the 1960s, as digital computers gained prominence, a small group of researchers began exploring a radically different approach to problem-solving. They asked a question that seemed almost absurd at the time: could **liquids**, with their natural ability to flow, mix, and adapt, be harnessed to perform computations? While most of the scientific community focused on silicon-based circuits, these pioneers were captivated by the potential of fluid dynamics to simulate complex systems.

The roots of liquid computing lay in **analog computation**, where physical systems—rather than digital logic—are used to solve mathematical equations. Water, with its dynamic behavior, became an obvious candidate. In one early experiment, a team constructed a model where water flowed through a network of pipes and tanks. By observing how the water distributed itself under varying conditions, researchers were able to simulate real-world processes like river systems and urban drainage networks.

One particularly famous demonstration involved a **maze filled with water**. Researchers injected a colored dye at the maze's entrance and watched as the dye spread, naturally finding the shortest path to the exit. This simple yet elegant setup visually solved an optimization problem, showing how fluid behavior could mirror mathematical computation. It was both mesmerizing and unexpectedly efficient.

Another pioneering experiment used tanks of water to model heat transfer. By carefully controlling inputs and outputs, scientists could visualize how heat dissipated in different materials, offering insights into thermodynamics that would have been computationally intensive for early digital machines. These experiments proved that liquid systems could mimic mathematical equations with remarkable accuracy.

Despite its promise, liquid computing faced significant challenges. Its systems were cumbersome and hard to scale, and their applications were limited to specific types of problems. Nevertheless, the quirky idea of using liquids as computational tools sparked a new wave of thinking about alternative approaches to problem-solving.

Importance and Impact

Liquid computers illustrated the power of **analog systems** to tackle problems that were dynamic, real-world, and difficult for early digital machines. By simulating physical processes directly, they provided insights into complex systems like weather patterns, fluid dynamics, and even traffic flow. These systems excelled in areas where the equations were too complex to solve directly, offering a faster and more intuitive approach.

One of the most significant legacies of liquid computing is its influence on **microfluidics**, a field that uses tiny channels of liquid to perform tasks traditionally handled by electronics. In healthcare, microfluidic devices have revolutionized diagnostics, enabling doctors to conduct rapid blood tests, detect pathogens, and study single-cell behaviors with unparalleled precision. These devices have roots in the same principles demonstrated by liquid computers decades earlier.

Liquid computing also inspired the development of **biomimicry algorithms**, where natural systems like fluid flow are used as models for artificial intelligence. For example, the way liquids navigate mazes has been used to design routing algorithms for traffic systems, logistics networks, and even internet data packets. These insights have found applications in fields as diverse as urban planning, supply chain management, and telecommunications.

The principles of liquid computing have even influenced modern **environmental modeling**. Researchers use similar methods to simulate the movement of pollutants through ecosystems or to study the impacts of climate change on water systems. While modern digital computers handle the heavy lifting, the conceptual framework provided by liquid systems remains invaluable.

Though liquid computers never replaced traditional systems, their quirky elegance and practical potential have ensured their lasting legacy. They represent a creative approach to computation, one that blends the natural behavior of fluids with mathematical problem-solving in ways that remain relevant today.

Chapter Summary

Liquid computers transformed the properties of fluids into tools for solving complex problems, from maze navigation to climate modeling. Their influence extends to modern microfluidics, environmental studies, and biomimicry algorithms. By demonstrating the potential of analog systems, liquid computing showcased how unconventional ideas can inspire profound innovations.

Next Chapter

As liquid systems revealed the potential of analog computation, another field was delving into the quantum realm to uncover the secrets of life itself. In the next chapter, we'll explore the emerging discipline of **quantum biology**, where the strange rules of physics help explain life's most fundamental processes.

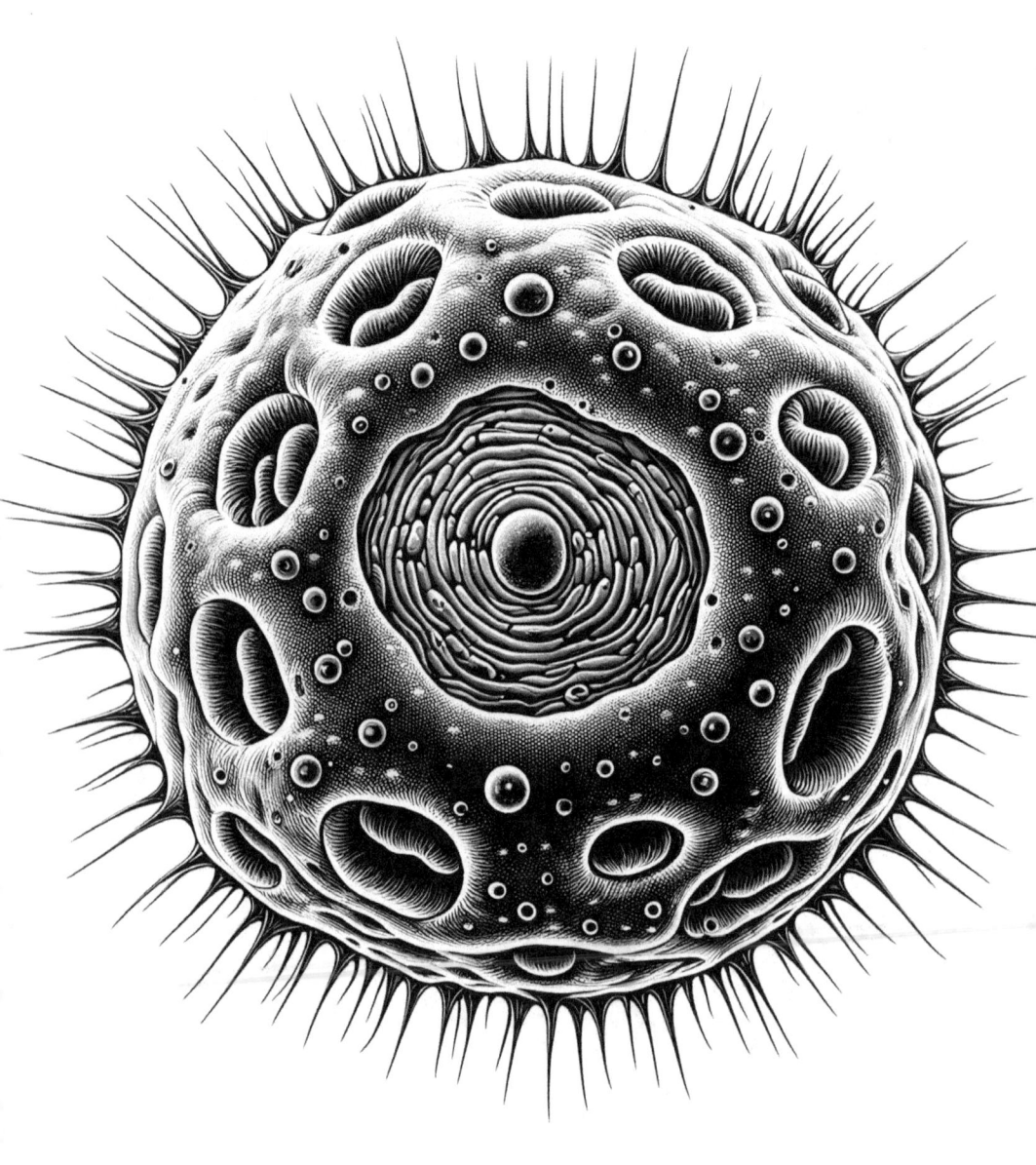

Quantum Biology
Life's Strange Physics Revealed

For much of the 20th century, the bizarre rules of quantum mechanics were considered the exclusive domain of physics. These principles, governing subatomic particles, seemed entirely disconnected from the macroscopic, messy world of biology. Yet, starting in the late 20th century, researchers began to uncover evidence that **quantum phenomena** were integral to life's most essential processes. This realization bridged two fields previously thought to be unrelated, creating the emerging discipline of **quantum biology**.

The first major breakthrough came from studies of **photosynthesis**, the process by which plants convert sunlight into chemical energy. Researchers discovered that the energy absorbed by chlorophyll did not move randomly through the plant. Instead, it used **quantum superposition** to explore all possible pathways simultaneously, settling on the most efficient route. This quantum efficiency, unmatched by human-made systems, allowed plants to harvest energy with extraordinary precision.

An equally strange discovery emerged from studies of **bird migration**. Certain migratory birds appeared to use a quantum-based compass, relying on entangled particles in their eyes to detect Earth's magnetic field. This phenomenon, still not fully understood, allowed birds to navigate vast distances with stunning accuracy, challenging conventional ideas about animal behavior and sensory biology.

Further research revealed that **quantum tunneling**, a process where particles bypass energy barriers, played a role in enzyme activity. Enzymes, the biological catalysts essential to life, used this quantum effect to speed up reactions far beyond what classical physics could explain. These findings hinted that quantum mechanics might underpin many of life's biochemical processes.

Despite initial skepticism, quantum biology gained traction as more evidence emerged. From the tiniest particles to entire ecosystems, this field revealed how the quantum world influenced life in ways both profound and surprising. By bridging biology and physics, quantum biology offered a new lens through which to understand the fundamental nature of life.

Importance and Impact

Quantum biology is transforming how we think about energy, medicine, and the nature of life itself. The discovery of quantum effects in photosynthesis has inspired efforts to create **artificial photosynthesis systems**, which could revolutionize renewable energy by mimicking the efficiency of plants. These innovations promise cleaner, more sustainable ways to power the future.

Quantum tunneling, another key phenomenon, has reshaped the field of **biochemistry**. Understanding how enzymes exploit quantum mechanics has led to breakthroughs in **drug design** and **catalysis**, allowing scientists to develop treatments for diseases with unprecedented precision. This research could pave the way for therapies targeting everything from cancer to aging-related disorders.

The implications of quantum biology extend beyond biology itself. By studying how nature processes information at the quantum level, researchers are advancing **quantum computing** and **artificial intelligence**. Insights from bird navigation and enzyme activity are inspiring smarter algorithms and more efficient technologies, blurring the line between natural and artificial systems.

Quantum biology also challenges long-held assumptions about the separation between physics and biology. By showing how quantum principles govern life, this field underscores the interconnectedness of scientific disciplines. It serves as a powerful reminder that the natural world operates on principles far more complex—and elegant—than we might have imagined.

Although the field is still young, its discoveries are already reshaping our understanding of both physics and biology. The study of quantum biology exemplifies the value of interdisciplinary research and the importance of embracing curiosity to uncover the hidden mechanisms of life.

Chapter Summary

Quantum biology reveals how quantum mechanics governs life's processes, from photosynthesis to bird migration and enzyme activity. This quirky, emerging field is inspiring innovations in energy, medicine, and technology, challenging traditional boundaries between scientific disciplines. It highlights the profound interconnectedness of natural systems and offers a glimpse into life's most fundamental mysteries.

Next Chapter: The Endless Horizon of Discovery

As we conclude this exploration of the nooks and crannies of science and technology, we are left with a tapestry of discoveries that illuminate the unexpected and brilliant paths of human curiosity.

From the accidental discovery of fullerenes, which shaped nanotechnology, to the cybernetic turtles that laid the foundation for robotics, we have seen how chance, creativity and perseverance turned strange ideas into transformative innovations. The bacterial battery showed us how nature's quirks could drive the future, while liquid computers demonstrated that unconventional thinking can unveil new ways to solve problems. We have witnessed the interplay between nature and technology, and how the two inspire each other in unexpected ways.

The main theme of this journey is the **power of curiosity-driven science**. Many of the breakthroughs we have explored began with simple, even playful questions: Can bacteria breathe metal? What happens when you shine a light through a nanostructure? Can fluids think? These questions, though unconventional, led to profound insights that transformed the world.

This book celebrates stories of **missteps**, **accidents**, and **serendipitous discoveries**, demonstrating that science thrives on the unpredictable. It reminds us that great discoveries often arise from exploring the unexplored and embracing the strange. The hidden mechanisms of the universe - whether in the quantum realm, in the depths of a riverbed or in the behavior of a particle of light - hold infinite surprises.

As science advances, new stories of unexpected discoveries are waiting to be written. This book may end here, but the journey of curiosity is far from over. The mysteries of the world, both natural and technological, remain immense and seductive, waiting for those who dare to ask strange questions and follow their trail.

Bibliography

General Science and Technology
- Asimov, Isaac. The Intelligent Man's Guide to Science. Basic Books, 1960.

- Bryson, Bill. A Short History of Nearly Everything. Broadway Books, 2003.

- Gleick, James. The Information: A History, a Theory, a Flood. Pantheon, 2011.

Quantum Biology
- Ball, Philip. Life on the Edge: The Coming of Age of Quantum Biology. HarperCollins, 2014.

- Al-Khalili, Jim, and Johnjoe McFadden. Quantum Biology: How Quantum Theory and Biology Work Together to Explain Life's Biggest Questions. Bantam Press, 2014.

- Lambert, Neill, et al. "Quantum Biology." Nature Physics, vol. 9, no. 1, 2013, pp. 10–18.

Microbial Fuel Cells and Bioremediation
- Logan, Bruce E. Microbial Fuel Cells. Wiley-Interscience, 2008.

- Rabaey, Korneel, and Willy Verstraete. "Microbial Fuel Cells: Novel Biotechnology for Energy Generation." Trends in Biotechnology, vol. 23, no. 6, 2005, pp. 291–298.

- Lovley, Derek R. "Electrogenic Microbes for Sustainable Power." Nature Reviews Microbiology, vol. 4, no. 7, 2006, pp. 497–508.

Nanotechnology and Fullerenes
- Kroto, Harry. Fullerenes: Past, Present, and Future. World Scientific, 1995.

- Sargent, Edward H. The Dance of Molecules: How Nanotechnology Is Changing Our Lives. Thunder's Mouth Press, 2005.

- Drexler, Eric. Engines of Creation: The Coming Era of Nanotechnology. Anchor Books, 1986.

Analog and Liquid Computing
- Bennett, Charles H. "The Thermodynamics of Computation—A Review." International Journal of Theoretical Physics, vol. 21, no. 12, 1982, pp. 905–940.

- Turing, Alan M. Collected Works of A.M. Turing: Mechanical Intelligence. Elsevier, 1992.

- Cardona, Manuel. "Fluid Computation Models: Exploring Analog Solutions in Engineering." Fluid Mechanics Today, vol. 12, 1997.

Bionic and Robotic Innovations
- Walter, William Grey. The Living Brain. W.W. Norton & Company, 1953.

- Brooks, Rodney. Flesh and Machines: How Robots Will Change Us. Pantheon, 2002.

- Sharkey, Noel, et al. Living Machines: A Primer on Biomimetic and Biohybrid Systems. Springer, 2012.

Biological Systems and Photosynthesis
- Govindjee, and David Krogmann. Discoveries in Photosynthesis. Springer, 2005.

- Blankenship, Robert E. Molecular Mechanisms of Photosynthesis. Wiley-Blackwell, 2002.

Cybernetics and Early Robotics
- Wiener, Norbert. Cybernetics: Or Control and Communication in the Animal and the Machine. MIT Press, 1948.

- Sharkey, Noel. "Cybernetic Organisms: The First Robotic Animals." Historical Robotics Journal, vol. 22, no. 3, 1999.

Cryptography and Codebreaking
- Kahn, David. The Codebreakers: The Comprehensive History of Secret Communication from Ancient Times to the Internet. Scribner, 1996.

- Navajo Nation Museum. Navajo Code Talkers: Their Contribution to WWII. Navajo Nation Press, 2010.

General References on Accidents and Serendipity in Science
- Roberts, Royston. Serendipity: Accidental Discoveries in Science. Wiley, 1989.

- Johnson, Steven. Where Good Ideas Come From: The Natural History of Innovation. Riverhead Books, 2010.

- Firestein, Stuart. Failure: Why Science Is So Successful. Oxford University Press, 2015.

Final note of thanks

To my insatiably curious readers,

Thank you for embarking with me on this journey through muddy waterways, swirling liquid computers and the quantum mysteries of life.

Your curiosity, your thirst for the unseen and the unexpected, is the spirit that fuels discovery. It is what transforms the ordinary into the extraordinary and keeps the flame of wonder alive in a world often too busy to realize the beauty of its hidden corners.

This book is as much his as it is mine. It is a celebration of those who refuse to settle for easy answers, who ask strange questions, and who dare to see the world not only as it is, but as it could be. If these pages have inspired you or encouraged you to ask a peculiar question, together we have achieved something wonderful.

Stay curious, be brave, and remember: the next great discovery could begin with your next question.

With gratitude and wonder,

O. Bennet

About the author:

O. Bennet

With a degree in Chemistry with a specialization in Polymers, and additional training in Computer Systems Administration and Programming, this author brings a blend of scientific and technological knowledge. His passion for learning and discovering new advances in science and technology fuels his constant curiosity unleashed in these fields.

Driven by his love of innovation and progress, he now shares his knowledge with readers through thought-provoking publications.

In them, he delves into the most fascinating marvels of the last decade, offering accessible and entertaining commentary on the discoveries that are shaping our incredible future.

Through his reading, he opens the door to ideas that will inspire readers to explore, learn and be captivated by the ever-evolving world of science and technology.

Disclaimer

Certain brand, product, and company names are referenced in this book for strictly educational and historical purposes.

These references are used solely to inform readers about major advances in science and technology and are not intended as an endorsement or for commercial use.

All trademarks are the property of their respective owners and do not imply any association or affiliation with them.

Copyright © 2024 Ely & Oly books
All rights reserved.